AF540666

Advances in Aquatic Ecology

Advances in Aquatic Ecology

Abhiraj Jain

RANDOM PUBLICATIONS
NEW DELHI - 110 002 (INDIA)

Advances in Aquatic Ecology

ISBN 978-93-51112-83-9

Published in 2014 in India by

RANDOM PUBLICATIONS

4376-A/4B, Gali Murari Lal, Ansari Road
New Delhi-110 002
Phone: +9111-43580356, 23289044
E-mail: randomexports@gmail.com; sales@randompublications.com; info@randompublications.com

Reprinted 2021

Type Setting by: Friends Media, Delhi-110089
Digitally Printed at: Replika Press Pvt. Ltd.

Preface

Aquatic ecology is an extraordinarily broad and interesting field. It investigates the interplay between aquatic organisms and their physical, chemical, and biological environment. Aquatic ecology encompasses all freshwater and marine ecosystems, including streams, rivers, lakes, wetlands, coastal environments, and the vast expanses of the open ocean. Aquatic ecology studies a wide diversity of different organisms, ranging from tiny bacteria to large whales, facing a myriad of different processes such as biogeochemical cycles, genetic differentiation, and climate change. Fundamental research in aquatic ecology adds new discoveries almost every day. Applied research makes major contributions to biotechnology, fisheries, water management, nature conservation, and environmental policy. Reassessments and syntheses in aquatic ecology are stimulating to the discipline as a whole, as well as enormously useful to students and researchers in ecological sciences. Marine ecosystems cover approximately 71% of the Earth's surface and contain approximately 97% of the planet's water. They generate 32% of the world's net primary production. They are distinguished from freshwater ecosystems by the presence of dissolved compounds, especially salts, in the water. Approximately 85% of the dissolved materials in seawater are sodium andchlorine. Seawater has an average salinity of 35 parts per thousand (ppt) of water. Actual salinity varies among different marine ecosystems. Marine ecosystems can be divided into many zones depending upon water depth and shoreline features. The oceanic zone is the vast open part of the ocean where animals such as whales, sharks, and tuna live. The benthic zone consists of substrates below water where many invertebrates live. The intertidal zone is the area between high and low tides; in this figure it is termed the littoral zone. Other near-shore (neritic) zones can include estuaries, salt marshes, coral reefs, lagoons and mangrove swamps. In the deep water, hydrothermal vents may occur where chemosynthetic sulfur bacteria form the base of the food web. Classes of organisms found in marine ecosystems include brown algae, dinoflagellates, corals, cephalopods, echinoderms, and sharks. Fishes caught in marine ecosystems are the biggest source of commercial foods obtained from wild populations. Environmental problems concerning marine ecosystems include unsustainable exploitation of marine resources (for example overfishing of certain species), marine pollution, climate change, and building on coastal areas. Aquatic ecosystems perform many important environmental functions. For example, they recycle nutrients, purify water, attenuate floods, recharge ground water and provide habitats for wildlife. Aquatic ecosystems are also used for human recreation, and are very important to the tourism industry, especially in coastal regions.

The health of an aquatic ecosystem is degraded when the ecosystem's ability to absorb a stress has been exceeded. A stress on an aquatic ecosystem can be a result of physical, chemical or biological alterations of the environment. Physical alterations include changes in water temperature, water flow and light availability. Chemical alterations include changes in the loading rates of biostimulatory nutrients, oxygen consuming materials, and toxins. Biological alterations include over-harvesting of commercial species and the introduction of exotic species. Human populations can impose excessive stresses on aquatic ecosystems. There are many examples of excessive stresses with negative consequences. Consider three. The environmental history of the Great Lakes of North America illustrates this problem, particularly how multiple stresses, such as water pollution, over-harvesting and invasive species can combine. The Norfolk Broadlands in England illustrate similar decline with pollution and invasive species. Lake Pontchartrain along the Gulf of Mexico illustrates the negative effects of different stresses including levee construction, logging of swamps, invasive species and salt water intrusion.

The contents of the book will help scientists and academics to examine, modify and improve of this subject.

—Abhiraj Jain

Contents

1

Introduction

Analysis of Pleistocene River Offsets

The Kisik river system has a simple morphology, its upstream part being offset from the Kisik-Aksu outlet gorge by c. 4.5 km. Its only significant tributary near the Gölbasi-Türkoglu Fault, the Cinarcik, is offset by c. 2 km further and flows for this distance along the linear valley along the fault before joining the Kisik. The headwaters of the Cinarcik on the Turkish side of the fault are offset by c. 4.5 km from those of a tributary of the Karatas River at [CB 6040 6560], near Soku on the Arabian side. Thus, restoring 4.5 km of slip on the Gölbasi-Türkoglu Fault juxtaposes both these headwater reaches with counterparts across the fault, the latter forming the upper reach of the now truncated Soku Karatas river.

This brief analysis can lead to the following deductions. If it is assumed that the landscape had relatively low relief in the Early Pleistocene, such that rivers were not yet entrenched into gorges, then both the Kisik and Cmarcik were able to flow across the Gölbasi-Türkoglu Fault with no deflection. At the start of faster regional uplift caused by climate change during MIS 22, the Kisik began to incise to try to maintain an equilibrium longitudinal profile. Being a relatively large river, with considerable erosional power at times of high seasonal flow, it was able to do this. It thus became 'locked' into its own gorge, which has since become progressively offset by left-lateral slip on the Gölbasi-Türkoglu Fault.

In contrast, the smaller Cinarcik with less erosional power, became deflected along the fault line to form a tributary of the Kisik, thus truncating its former lower reach on the Arabian side of the fault.

We next consider whether this view, that the Golbasi-Turkoglu Fault has slipped by c. 4.5 km since MIS 22, is supported by the evidence from the Koca and Gök river systems. For the Koca system, indicates that, after 4.5 km of left-lateral slip is restored, the points where the Sincer and Velikler rivers reach the fault line on its Turkish side are juxtaposed opposite what are now truncated headwaters of affluents of the Soku Karatas River and of another Karatas River, which flows separately into the Aksu past the village of Payamhbag. The Kaykirth tributary of the Koca is aligned opposite the truncated headwaters of a more westerly affluent of the Payamlibag Karatas River, which is now a right-bank tributary of the Koca, joining it downstream of the fault.

The Koca itself aligns with the point where the combined Sincer and Velikler now leave the fault line on its Arabian side. These correlations imply that a complex sequence of drainage diversions affected what are now tributaries of the Koca system during the Mid-Late Pleistocene. For instance, the modern outlet gorge of the Koca away from the fault line is predicted to have aligned in MIS 22 with the gulleys in the karstified outcrop of Midyat Group limestone on the Turkish side of the fault forming Ahlankavagi Tepe near Kocalar. The modern offset of the Koca, upstream of this present outlet, from [CB 5020 6030] to [CB 5270 6160] is c. 2.8 km, suggesting that it adopted this route around 870 ka × 2.8/4.5 or c. 540 ka, indicating MIS 14. It had presumably maintained its course through its original outlet until that time, before its upper reaches were captured by its present outlet gorge, which appears to have been the original drainage outlet for the Ahlankavagi Tepe area.

The Velikler, Sincer and Kaykirth were presumably diverted into the linear valley along the Gölbasi-Türkoglu Fault immediately after the increase in regional uplift rates, their outlet from then on having been the gorge that continues to be occupied by the Sincer, upstream of its modern confluence with the Koca on the Arabian side of the fault. For much of the Mid-Pleistocene, these rivers would thus have been offset to the right across this linear valley, this pattern of offsets having since been gradually cancelled out by the progressive left-lateral slip on the fault.

For what is now the Gök system, restoring 4.5 km of leftlateral slip aligns each of the modern left-bank tributaries of the Gök, from the Turkish side of the Gölbasi-Türkoglu Fault, with an initial counterpart on the Arabian side. Thus, the Ahlankavagi Tepe gulleys align with the modern outlet gorge of the Koca, the Mezayok aligns with the Aslanbükü, the Büyük with the Kuru (i-v), the Cati with the Tavsan Tepe dry valley (j-w), the Cobanpmar with the modern outlet gorge of the Gök (k-x), the Karaagaç with the Sazh (1-y), and the Alanyolu with the Kara. It thus appears that the Cobanpmar has maintained its original geometry, with progressive deflection to the left as a result of slip on the Gölbasi-Turkoglu Fault; the Alanyolu and Karaagaç appear to have been promptly captured by and deflected into this system following the increase in regional uplift rates.

In contrast, the substantial incision evident in what is now the Tavsan Tepe dry valley suggests that, for a time, it formed the outlet for the Büyük and/or the Çati. However, such a geometry can have been maintained until no later than c. 600 ka (MIS 15), by which time the Büyük is predicted to have become juxtaposed opposite the modern Gok outlet gorge. This restoration suggests that the Mezayok, which is now a tributary of the Gök, was, for a time after MIS 22, a tributary of the Koca, joining it at the outlet from what is now the dry valley ENE of the Kartashk col. The c. 0.5 km length of this dry valley suggests that it was occupied by the Mezayok for around 100 ka, before this river was captured by the Gök system.

The consistency of these c. 4.5 km river offsets suggests strongly that the stated hypothesis is tenable. The 13 individual measurements of offsets between piercing points listed give a mean value of 4.44 ± 0.06 km (±2 <3). The left-lateral slip rate on the Gölbasi-Türkoglu Fault, time-averaged since MIS 22 (870 ka), is thus estimated as 5.10 ±0.07 mm a-1. As already noted, the Westaway (2004a) kinematic model predicts Turkish-Arabian relative motion at 8.0mm a^sup -1^ towards S48°W in the vicinity of Gölbasi. The above estimate that 5.1 mm a-1 of this motion occurs on the Gölbasi-Turkoglu Fault leaves an additional 2.9 mm a -1 unaccounted for, which can be presumed to be partitioned between the Sürgü Fault and the array of NNE-SSW-striking faults south of the Gölbasi-Türkoglu.

The combined total slip on the Gölbasi-Türkoglu Fault and these NNE-SSW-striking faults was estimated as 33 km by Westaway & Arger (1996), from offsets J-M and Q-T, a deduction that is supported by the present study. To make any further refinement of the regional kinematics, an independent estimate of the total slip on the Gölbasi-Türkoglu Fault is needed. Earlier discussion established that, because the precise locations of the piercing points that were used (after Tolun & Erentöz 1962) cannot be unambiguously established, the reasoning that led to the Westaway & Arger (1996) estimate of 16 km needs to be re-examined. As Westaway & Arger (1996) noted, the Ahlankavagi

Tepe-Kozdag piercing point adjoined the site where, at its time of formation, the GölbasiTürkoglu Fault splayed from the zone of NNE-SSW-striking left-lateral faulting farther south. To constrain the slip on just the Gölbasi-Türkoglu Fault one must thus investigate sites to the west of these points.

The possibility of differential erosion between different parts of any strike-slip fault in SE Turkey makes it difficult to establish exact matches between piercing points, and also means that apparent offsets between outcrop boundaries do not necessarily equal their true strike-slip offset. However, four general forms of evidence can provide piercing points to constrain the total slip on the Gölbasi-Türkoglu Fault: the unconformities between the ophiolite and overlying sediments at the eastern margin of the ophiolite; the relief at the western margin of the ophiolite; the highest summits within the ophiolite; and the locations of the highest ridges within the ophiolite. The preponderance of evidence suggests c. 19 km of offset, indicating the total slip on the Gölbasi-Türkoglu Fault.

The most reliable geomorphological indicator seems to be the match between the ridge of ophiolite forming Köroglu Tepe on the Turkish side of the Gölbasi-Türkoglu Fault NE of Türkoglu against the highest topography on the Arabian side of the Gölbasi-Türkoglu Fault south of Göl Alani. Matching of the highest topography on the Turkish side of the GölbasiTürkoglu Fault north of Göl Alani against that on the Arabian side around Karaagac Tepe provides a similar estimate, as does matching the westernmost evidence of the presence of ophiolite near Turkoglu on the Turkish side of the Gölbasi-Türkoglu Fault and near Çigli on the Arabian side. Matching the contact between ophiolite and Besni Formation in the Gok gorge near Sürülü on the Turkish side of the Gölbasi-Türkoglu Fault against its counterpart in the Kisik gorge on the Arabian side near Soku gives a similar estimate. Matching the eastern margins of the ophiolite outside these gorges underestimates the other values by c. 1 km.

A possible cause of this mismatch is faster erosion of the feather edge of the Besni Formation adjacent to the Gök gorge, possibly as a result of higher rainfall in this higher-altitude locality; if the Besni Formation persisted farther west in this locality the offset measurement would be greater. Alternatively, the ophiolite may persist farther east on the Arabian side of the Gölbasi-Türkoglu Fault, leading to the alternative c. 19 km slip estimate W-Y'. Dividing this 19 km estimate for the total slip on the GölbasiTürkoglu Fault into the 5.10 ± 0.07 mm a -1 slip rate deduced earlier gives an estimated age (assuming that the same slip rate has been maintained throughout) for the Gölbasi-Türkoglu Fault of 3.73 ± 0.05 Ma. Conversely, dividing the total of 33 km of slip across all left-lateral fault strands in the Gölbasi area by the 8.0 mm a -1 rate of Turkish-Arabian relative motion (from West-away 2004a) gives a higher age estimate, of 4.13 Ma, and implies that 14 km of this 33 km of relative motion was taken up on NNE-SSW-striking faults south of Golbasi.

A possible explanation for this discrepancy, after Westaway (2004a), is that part of the slip on these NNE-SSW-striking faults predates the modern East Anatolian Fault Zone system. The total of Turkish-Arabian relative motion expected since 3.73 ± 0.05 Ma can be estimated as 3.73 ± 0.05 Ma × 8.0 mm a"' or 29.8 ± 0.4 km. Neglecting for the time being, for convenience of calculation, any slip on the Sürgü Fault on this time scale, one can infer that a minimum of c. 11 km of the slip on these NNE-SSW-striking faults has accompanied the modern geometry of the East Anatolian Fault Zone; the rest (up to c. 3 km) may thus predate it. Alternatively, the 8.0 mm a-1 rate of Turkish-Arabian relative motion at Gölbasi, from Westaway (2004a), which has formed the basis for these calculations (and which was derived from GPS data, from McClusky et al. (2000), not geological evidence) may slightly underestimate the overall time-averaged rate.

Adjusting this rate upward to 8.85 ± 0.12 mm a-1 would predict 33 km of total slip since 3.73 ± 0.05 Ma, thus accounting for the geological evidence from this area, without any need for any of this slip to predate the Mid-Pliocene initiation of the East Anatolian Fault Zone system. More detailed analysis of this issue will require quantitative analysis of the array of NNE-SSW-striking faults south of Gölbasi, and so is beyond the scope of the present study.

However, adding the c. 4 km estimate of left-lateral slip on the Sürgü Fault (Westaway 2004a) to the c. 33 km on the Gölbasi-Turkoglu Fault gives c. 37 km, suggesting an overall local rate of Turkish-Arabian relative motion of 9.92 ± 0.13 mm a-1 since 3.73 ± 0.05 Ma. Assuming the same position for the pole of rotation as was deduced by Westaway (2004a), the Turkish-Arabian Euler vector adjusts to 0.89 ± 0.01° Ma^sup -1^ about 33.4°N, 42.3°. It now appears that increases in uplift rates following the onset of 100 ka Milankovitch climate cyclicity in the late Early Pleistocene (MIS 22) are a widespread phenomenon worldwide, which can be attributed to coupling between surface processes and the lower-crustal flow that is induced in response in order to maintain isostatic equilibrium. The analysis in the present sudy indicates that the onset of this combination of coupled processes led to a complex series of drainage adjustments in the study region, as rivers became progressively entrenched into what had previously been a low-relief landscape with linear drainage. However, this sequence of adjustments seems to have been completed in this study region no later than MIS 14, since when the new, much more dendritic, drainage geometry that developed has remained stable. As previously inferred, this uplift has nothing directly to do with the left-lateral faulting, and would be occurring even if this region were not within a plate boundary zone.

Regarding rates and amounts of uplift during the Mid- to Late Pleistocene, it is evident that, with minor exceptions, river gorges that have been incised on this time scale are c. 100-120 m deep. This includes the gorge of the Aksu trunk channel throughout most of its length, and many of its tributary gorges. The principal exception is in the headwaters of tributary gorges, where they have little erosional power and so have been unable to incise at the same pace as the regional uplift, as is also widely observed in many other regions. Restoring this incision indicates that this study region had much lower relief in the Early Pleistocene than at present. The principal relief that already existed related to the earlier folding of the erosion-resistant Midyat Group limestone, notably as mountains such as Sakarkaya Dag and Çatal Dag, as well as in smaller folded inliers of it such as Atatar Kayasi. Other relief dating from this time related to the roughness of the land surface in the ophiolite.

In some places the ophiolite has effectively 'armoured' the flanks of adjoining deposits that would otherwise have been more easily erodable, thus protecting them from erosion, for instance preserving extensive deposits of the Besni and Germav Formations in much of the Gök system on the Turkish side of the Gölbasi-Türkoglu Fault. However, elsewhere, where no such 'armouring' effect has been possible, the more easily erodable deposits have largely been removed from the region, a notable example being provided by the Koca river west of Kocadere which appears to have removed the deposits of the Selmo Formation from the northern flank of its gorge almost as fast as it has itself incised, creating a low-relief landscape just above modern river level. As already noted, the largest river in the Arabian Platform of SE Turkey, the Euphrates (Firat) has a well-developed staircase of Pleistocene terraces, notably around Birecik.

This river flows c. 50-60 km east of the present study region; it is thus expected to indicate the same uplift history as the smaller rivers investigated in the present study. The terrace that is now thought to date from MIS 22 is found c. 110 m above present river level, indicating essentially the same amount of incision on this time scale as is observed in the river systems investigated in the present study, even though the latter are much smaller. We thus interpret this incision as a good proxy for regional uplift by a near-uniform distance across this region on this time scale.

The principal evidence for lateral variations in regional uplift in the present study region is provided by the difference in altitude, by c. 400-500 m, between the outcrops north of the Gölbasi-Türkoglu Fault near Türkoglu and south of the Gölbasi-Türkoglu Fault in the vicinity of Göl Alani. We presume that this differential effect is the net isostatic result, since the Pliocene, of the sediment loading that has occurred in the Plio-Pleistocene depocentres that adjoin the Gölbasi-Türkoglu Fault around Türkoglu and Kahraman Maras. However, detailed calculation of this effect, for instance using the physics-based modelling approach of Westaway, is beyond the scope of this study.

It is becoming increasingly clear that increases in uplift rates, revealed by fluvial incision, resulting from global climate change in the late Early Pleistocene, are widely observed, being well-documented in temperate latitudes and also evident within the tropics, not just in Europe where this effect was first identified. It follows that detailed analysis of the fluvial geomorphology along many other active strike-slip faults worldwide may well provide analogous datasets to that developed in the present study, which can give the slip rate time-averaged since MIS 22. Such estimates are likely to be extremely useful, for instance in earthquake hazard assessment, for testing regional kinematic models, and for comparison with shorter time scale datasets from trenching and GPS. The basis of a powerful new method is thus apparent.

In the Arabian Platform of SE Turkey, fluvial incision since the late Early Pleistocene has accompanied regional surface uplift, as the isostatic response to regional erosion. This incision, typically by c. 110 ± 10 m, starting in or around MIS 22 at 870 ka. has progressively 'locked' rivers into their gorges in landscape that formerly had much lower relief. We use this effect to estimate 4.44 ± 0.06 km of left-lateral slip on this time scale on the Gölbasi-Türkoglu Fault from offset river gorges, giving a slip rate of 5.10 ± 0.07 mm a^{-1}. Piercing points indicate that this fault has slipped a total of 19km, making its age 3.73 ± 0.05 Ma. A total of 33 km of relative motion between the Turkish and Arabian plates is documented on this time scale in the vicinity of

Golbasi, at an overall time-averaged rate of 8.85 ± 0.12 mm a-1. This analysis indicates the basis for a new method for estimating slip rates, time-averaged since the late Early Pleistocene, which is potentially applicable to strike-slip fault zones worldwide.

The Legend of St Cuthbert's Beads

St Cuthbert's Beads are 300-million year-old fossil crinoid columnals found on Lindisfarne, the Holy Island. The legend associating them with St Cuthbert, either that he manufactured the beads or that they were part of his rosary, originated as early as 1671 and spread through the British palaeontological literature. The source of the legend may be small columnals scattered among shells and pebbles on the beach below the village.

These weather out of glacial till. Alternatively, the origin may be related to the onset of limestone quarrying on the north side of the island beginning in 1344. Quarrymen finding the columnals may have carried them back to villagers who associated them with St Cuthbert. We became interested in the tale of St Cuthbert's beads and the Holy Island of Lindisfarne, because both of us have devoted a large part of our scientific research to the study of fossil crinoids. We first learned that St Cuthbert's beads referred to the individual columnals of fossil crinoids a good many years ago.

Despite the fact that reference to this old term is scattered through the palaeontological literature and that it is well known in British folklore, the origin of the legend is poorly constrained. We began actively investigating it when one of us (G. L.) first visited Lindisfarne in 1995. Since then, G. L. has travelled to the Holy Island in 1998 and 1999, and both of us visited the island in May 2000, to study the local geology. We wanted to ascertain how and when the legend originated and the geological and palaeontological basis for the legend. Summaries of the geography, natural history and human history of the island may be found in Cartwright and Cartwright (1976), Galliers (1970), and O'Sullivan and Young (1995). Crinoids are a class of the phylum Echinodermata, which includes living starfishes, sand dollars and sea urchins, as well as approximately six hundred living species of crinoids. Most fossil crinoids have the main visceral body raised above the sea floor by a stem, also called a stalk or column. This consists of a stack of individually secreted limy skeletal pieces called columnals, which are the hallmark of St Cuthbert's beads. Columnals are bound together in life by ligaments. However, when the animal dies, the ligaments decay and the column falls apart. Each columnal typically has a circular or pentagonal outline, although the circular ones are more common.

Each columnal has a hole through its centre, which may be circular or pentagonal in outline. This hole is called a lumen, and in life, it forms a hollow tube through the centre of the column. Within this tube, there is an extension of the coelomic sac, which contains a nerve cord. In fossil crinoids, the lumen is commonly filled with rock matrix, but if this matrix weathers out, as it commonly does, the individual columnal is very bead-like in appearance.

Cuthbert and Lindisfarne

The life of St Cuthbert is very well known, based on the early Life of Cuthbert written by the Venerable Bede in the eighth century. Although Cuthbert was Bishop of Lindisfarne for only one year (685-686 AD), he achieved early religious fame because of his sanctity and austere life, and became the focus of an early religious cult and community. Viking raids between 793 and 875 destroyed the monastery on Lindisfarne and finally forced the monks to abandon the island, taking Cuthbert's body with them.

From 875, the island was apparently uninhabited for more than two hundred years, but by at least 1122 a new church had been built on the island, and a Benedictine priory was established c. 1160. Priory records still exist that mention limestone quarrying in 1344, presumably on a small scale and mainly for building stones, or for burning lime for plaster. The limestone outcrops along the northern coast of the island undoubtedly provided the rock for this early quarrying operation.

Palaeontological and Literary Citations

The earliest written reference to St Cuthbert's beads that we have been able to find is by John Ray in 1671. He visited the island and must have heard reference to the beads from the local villagers. In his itineraries, Ray stated: "July the 22nd we rode from Cheviot, or rather Waller or Wooler, to the Holy Island, nine miles, where we gathered, on the sea shore under the town, those stones which they call St Cuthbert's beads, which are nothing else but a sort of entrochi." While these remarks were not published until much later, Ray did publish a note about the beads in 1673: "II. Those they call S. Cuthbert's beads are found on the Western Shore of the Holy Island".

In the same year, Martin Lister published the first account of British fossil crinoids, which he thought might be "plants petrified." In the lead paragraph of this chapter Lister said: "In this chapter I send you an account of some of the Parts of certain Stones figured like Plants, which Agricola calls Trochitae, and the compound ones Entrochi, we in English, St Cutberds beads".

Lister and Ray were correspondents and Ray may have transmitted the Cuthbert legend to Lister and other British scientists after his visit to the Holy Island. Any explanation of the origin of the legend must take into account the fact that the legend was clearly firmly established by the early date of 1673. Fifteen years later, Robert Plot stated: "Many of these being perforated some with a round, others with foliated or asterial inlets of 6 or 7 points, anciently when found single or but double or treble ... they were strung like beads, particularly by St Cuthbert, which gave occasion to their other name of St Cuthbert's beads". Plot surely inferred that Cuthbert would have strung the beads together to form a rosary? Many other brief references to St Cuthbert's beads that occur in the literature maintain the association between fossil crinoid columnals, St Cuthbert, and Lindisfarne. The earliest reference to St Cuthbert actually making the beads is by Francis Grose, who said that "according to the vulgar belief, he often comes thither in the night, and sitting upon a certain rock uses another as his anvil, on which he forges his beads."

The earliest illustration of crinoid columnals specifically from the Holy Island that we have been able to find is in a 1792 letter by "R. W." in The Gentleman's Magazine. In this long letter, R. W. noted that the beads "are found on the beach to the West of this island, but on no other part of the beach... On the North East side of the island, however, there is a large track of limestone, which abounds with these concretions". "D. H.," commenting on this letter, noted that "St Cuthbert's beads, are picked up among the rocks [of Lindisfarne] by the children who sell them to travellers". Sale of the beads at gift and souvenir shops on the island continued until about ten years ago.

Among modern references to the beads, M. G. Bassett summarised the St Cuthbert story and mentioned that St Cuthbert's name is also associated with the Whitby "snakestones," fossil ammonoid cephalopods. According to Basset, snakestones were specifically allied to the Holy Abbess Hilda, a contemporary of Cuthbert. Cuthbert was present at the Synod of Whitby in 664, and is said to have put a curse on the fossil ammonoids so that they lack heads (Bassett 1982). Many of the citations for St Cuthbert's beads in both the palaeontological and the popular literature of Great Britain are based on Sir Walter Scott's epic poem, Marmion (1808):

But fain Saint Hilda's nuns would learn
If on a rock, by Lindisfarne,
St Cuthbert sits, and toils to frame
The sea-bom beads that bear his name:
Such tales had Whitby's fishers told,
And said they might his shape behold,
And hear his anvil sound;
A deadened clang—a huge dim form,
Seen but, and heard, when gathering storm
And night were closing round.
But this, as tale of idle fame,
The nuns of Lindisfarne disclaim (canto 2, verse 16).

In addition to this specific reference to the beads, Scott's poem includes many references to the Holy Island and to Cuthbert. In summary, Cuthbert's name was associated with fossil crinoid columnals by the Lindisfarne islanders as early as 1671. The legend that Cuthbert manufactured the beads on a stone anvil was in existence as early as 1783, and the use of the beads as a rosary was recognised by 1686.

Palaeontology and Geology of the Holy Island

The overall geology of the Holy Island is relatively simple and it has been explained clearly by Galliers (1970) and Robson (1981). A series of marine limestones that are part of the Carboniferous Middle Limestone Group form the base of the island.

These crop out along the northern coast foreshore with a gentle dip to the east, so that the oldest limestone, the Eelwell, was formerly exposed near the western edge of the island and the youngest, the Sandbanks Limestone, forms the cliffs of Castlehead Rocks to the east. Between these units is a middle limestone, the Acre, which forms Snipe Point and the extensive Back Skerrs platform on the western side of that point. Interbedded between these three limestones are sequences of shales, fine-grained sandstones and thin coals that crop out poorly, and that are known in part from drilled boreholes.

Their stratigraphic intervals are represented along the northern coast by sandy beaches between the rocky points formed by each of the limestones. Over the top of these Palaeozoic rocks, there are widespread Pleistocene gravels and glacial tills as well as modern stable and active sand dunes, especially in the far western and northern parts of the island.

Holy Island Quarries

The only visible quarry on the island today is the large one at Nessend, just inland from Castlehead Rocks. This quarry was operated during the nineteenth century until approximately 1883 (Jermy 1992). It is still possible to see the wagon entrance to the quarry at the north-eastern corner and remnants of haulage roads to lime kilns.

On a geomorphological map of the island, Galliers showed two additional quarries. These were both south-west of Snipe Point, and both are indicated as being largely surrounded by sand dunes. There is no evidence of quarrying activity inshore from the coast anywhere along this part of the island. The more north-easterly of the two quarries is in the position of the medieval settlement at Green Shiel, which was investigated by O'Sullivan and Young. The Newcastle geomorphology students apparently mistakenly thought that the piles of stone that mark this archaeological site were signs of quarrying activity. The second quarry site just to the south-west of the one discussed earlier occupies a small basin among the dunes. In this basin, there are the ruins of a bothy. There is no indication of quarrying.

An indication of early working of limestone in the Green Shiel area is a remnant of an old haulage road that leads directly to the Back Skerrs limestone platform where there is virtually no blockage of the foreshore by sand dunes. The haulage road is about ten feet wide and was raised above the adjacent ground by use of cobbles from the nearby medieval settlement ruins. This may be part of a tramway shown on an Admiralty chart of about 1855 reproduced by Jermy (1992, 23). A horse-drawn cart could have moved along this road directly onto the limestone platform where the stone has two sets of vertical joints at right angles to each other that break the beds into easily removable blocks.

Origin of the Legend

Authoritative studies of St Cuthbert and his later followers continuing up to 1200 do not mention the bead legend. Priory records of the time indicate that there was quarrying of limestone as early as 1344 (Jermy 1992). Therefore, it seems likely that the legend arose many centuries after Cuthbert, perhaps in the fourteenth century when limestone quarrying began. There are three possibilities as to the source of the columnals and the origin of the legend.

- Villagers may have noticed rare columnals on the foreshore below the village, which weathered from glacial till. We have found small columnals embedded in the till and weathered out among shells and pebbles from just west of Heaugh Hill to Chares End, a distance of half a mile. The columnals are not common and cannot be found without a fingertip search.
- A second scenario is that quarrying operations exposed easily weathered fossiliferous shales, mudstones, or fine-grained sandstones interbedded with limestones. Cast aside by the workers into spoil heaps, these rocks would have decomposed readily, releasing the columnals from rock matrix. These loose columnals may have had soft matrix removed from the central lumina and, thus, would have been very bead-like in appearance. The workers may have gathered them up and carried them back to the village, where they may have been sold to travellers or been used to make rosaries. Their ancient appearance may well have led local people to believe that they were several centuries old, perhaps from the time of Cuthbert. A modern analogue to this second possibility can be seen today on the floor of the nineteenth-century Nessend quarry. There are long rows of spoil rock—shales and fine-grained sandstones—cast aside by the quarrymen. In places, these are fossiliferous, and crinoid columnals weather out in profusion. Approximately half of the columnals have an open lumen and, thus, are ready to be strung as beads.
- A third possibility is that columnals were found in unmetamorphosed shales along the north shore. Crinoid columnals do weather free from such shale exposures, but these exposures are limited and would not provide an abundant source of columnals which would be expected to be readily noticed.

There are rare cross-sections of columnals in hard, dark, fine-grained limestones which crop out along the south shore. However, the columnals commonly weather out at various angles so they lose their bead-like appearance. The same is true for the more common columnals that occur in limestone beds on the north shore. Thus, we favour the possibility that the legend arose either from columnals weathered from glacial till along the south shore beaches, or from loose weathered columnals that were the result of ancient quarrying activities on the north shore.

Human Interaction with Disturbance

Humans can have surprising and profound effects on how natural disturbances affect biotic communities. Human activities can change the scale of natural-disturbance events, increasing or decreasing the frequency or magnitude of some, and can create a directional rather than cyclic community change. By altering the scale of disturbance or intensifying the severity, humans can homogenise the landscape by reducing the number and variety of biotic elements. Even without the interaction of storms, inhabitants of small villages on Rio Lagartos and Celestun have adverse effects on plant communities, primarily mediated through the hydrological balance.

Early salt-gathering, which probably predated the Spanish conquest, required alteration of the barrier beaches and impoundment of waters of the lagoon. But water impoundment for salt gathering is currently intensifying for growing populations. On Rio Lagartos, this has already resulted in forest mortality from the unintentional flooding of inner-island forests with saltwater. Likewise, the construction of a new bridge from the mainland to Celestun has altered tidal patterns and created a zone of dead mangrove forest. Any human alteration of the terrain is likely to have an effect on the hydological balance. But the adverse effects of these changes are likely to be greatly exacerbated during severe storms.

A second bridge, built to connect Rio Lagartos with the mainland, illustrates the potential for destruction and unforeseen interaction of human structures with disturbance. The bridge trapped the storm surge forced into the lagoon by Hurricane Gilbert, which flooded and drowned large numbers of flamingo fledglings in their nests, although conservation of the species was a high priority. A series of potential changes for the barrier beaches can be expected with extensive development for tourism. Tourism can entail not only bridges but also buildings with foundations, roads, dikes, and buried pipes and sewer systems. The hydrological balance between saline water and freshwater is the component of the barrier-island system most vulnerable to this type of construction. The confined aquifer is the primary source of freshwater for the barrier-beach communities. If the confining layer is breached, a dramatic decrease in the thickness of the freshwater lens could result, with serious alteration of conditions for ecosystems.

Destructive effects of construction would be greatly intensified in the event of a major storm. Channels and construction would accelerate penetration of storm surges through and across the barrier peninsulas.

Seawater would be forced into the freshwater lens through access created by construction. Precipitation runoff from roads and parking areas could increase the severity of flooding. Alteration of vegetation communities would lead to further changes in a kind of positive feedback. For example, removal of mangrove forests fringing the coast and lagoon eliminates a physical barrier buffering interior natural communities and exposes them to increased wind and salt disturbance. Direct effects of human use of plant communities, even walking on dunes and beaches, would accelerate plant diminishment. With reduced plant cover that follows extended human use of a landscape, depauperate communities would be further decimated with each subsequent storm (Linhart 1980).

The first step in integrating an awareness of natural disturbance into planning for development is a description of the nature and the scale of disturbance. Disturbance regimes can be quantified by extent, frequency, and severity. Often, these can be approximated from existent information. For example, in the Yucatan, paths and frequencies of hurricanes have been known for more than a century, and although the actual effects on communities are not known, the shape of the disturbance regime can be predicted on average, much as hundred-year and five-hundred-year floods are. Second, the impact of the combination, even synergy, of a built environment and storms on various communities should be estimated to establish priority zones that should be protected from building and use.

Certain communities, such as the protective mangrove forests, can be considered keystone communities that, when removed, begin a cascade of destruction of other communities. There is a considerable body of evidence from hurricanes Gilbert and Andrew, for example, detailing the effects of this type of storm on both social and human-altered natural environments. Recent biogeographical research presents a convincing case that disturbance is common to many biological systems and is linked in time and space. Sociocultural changes put both human and natural communities at risk, a pattern nowhere clearer than in the burgeoning tourism on tropical coasts, where severe hurricanes are a historical reality.

Trends in ecological thought suggest that understanding human transformation of natural communities must include the dynamics of the natural world. Human designs must accommodate natural dynamics, including disturbances. Rational ecotourism development can be difficult to accomplish; indeed, though commonly viewed as a mode of conservation of biodiversity, nature tourism can also have adverse if not disastrous effects on biota. In developing the Yucatan coast for tourism, the design of structures, their placement, and their relationships to the delicate hydrological balance will strongly influence the sustainability of natural communities in the future. Although understanding the parameters of disturbance in a landscape tests human powers of prediction, the recognition that disturbance is one ecological component in the context of development may mitigate human-imposed effects on the natural environment.

The effects of collision between the Arabian and Eurasian plates, and the associated escape of the Anatolian microplate to the west, dominate the landscape of eastern Turkey and surrounding areas. A series of well-known major faults (Dewey et al. 1986; Bozkurt 2001) reflect both thrusting and strike-slip displacement, although recent seismicity (Orgulu et al. 2003) suggests that the latter is currently dominant. One consequence of this in eastern Anatolia is a series of fault-bounded basins representing localised extension induced by lateral shear and regional uplift (Dewey et al. 1986).

Extensive volcanism associated with these basins took place during the Neogene and has resulted in thick successions of igneous rocks, the upper surfaces of which are often planar, forming plateaux. Current volcanic activity is centred on crustal structures where movement is still occurring (Adiyaman et al. 1998), most notably in pull-apart basins bounded by active faults (Karakhanian et al. 2004). Much of the area has not been directly affected by volcanism since the late Tertiary, and has experienced landscape evolution in response to other controls (principally tectonic activity and climatic change). Research in eastern and southern Turkey suggests that the style of tectonism has varied through time, reflecting initial collision-related compression, then a more complex, spatially variable lithospheric response as the faults bordering the Anatolian microplate developed during the Pliocene.

Much of the existing work on this response has focused on the major tectonic lineations of the Northern Anatolian Fault Zone and East Anatolian Fault Zone, which meet near Karliova (e.g. Over et al. 2004b). In comparison, there has been limited consideration of the complexity and chronology of tectonic processes north and east of Karliova, in the 'East Anatolian Contractional Province' (Bozkurt 2001) despite the evident active faulting and uplift, and the historical record of damaging seismicity. This study focuses upon a small area in the East Anatolian Contractional Province, the Pasinler Basin

(39°58'N, 41°58'E), c. 40km east of the city of Erzurum. The approach combines field survey and air photograph interpretation with remotely sensed data. Pasinler is particularly suitable because of the range of clearly expressed and contrasting landforms present within a limited area. Moreover, this area is believed to be representative of the wider contractional province (Rust et al. 1999) and insights gained here may aid understanding of this relatively unstudied region, and this approach may be generally applicable to studies within similar tectonic regimes worldwide.

Although the basin is crossed by several faults, some areas appear not to have been directly affected by recent movements and thus have the potential to preserve features reflecting other palaeoenvironmental controls. The area is also seismically active, with a long historical and instrumental record of damaging earthquakes; for example, in 1924 and 1983 (Eyidogan et al. 1999). Improved understanding of the geomorphological expression of this activity should contribute to future research on longer-term hazard assessment.

The Pasinler Basin

The Pasinler Basin originated in the Miocene through extension, associated with regional uplift of the East Anatolian Accretionary Complex and perhaps induced by slab breakoff (Keskin 2003), with Eocene volcanic and sedimentary rocks being overlain by Miocene and Pliocene igneous materials that accumulated to thicknesses of several hundred metres (Keskin et al. 1998). The last significant igneous activity in the immediate vicinity of the Pasinler Basin is represented by lavas and ignimbrites produced after about 7.8 Ma that underlie the Erzurum-Kars Plateau. Younger volcanic rocks occur to the south and east (Keskin et al. 1998), suggesting a significant migration of the regional stress field since the Miocene (Koçyigit et al. 2001). During the Miocene and Pliocene, erosion of the Plateau sequence was associated with deposition of thick fluvial and lacustrine sequences and this denudation has continued through the Pleistocene and Holocene, with the basin floor occupied by extensive clastic sediments at least 80 m thick (Bozkus 1993). Low-magnitude seismicity (M <4) is frequent and the basin has been affected by a number of higher magnitude destructive events, most recently the Pasinler and Horasan earthquakes in AD 1924 and AD 1983, respectively (both M 6.8: Bayraktutan et al. 1986; Eyidogan et al. 1999). Seismicity is mainly associated with strike-slip movements along the NE-SW-trending North East Anatolian Fault zone and related structures, although some appears to be linked to thrust faulting (Eyigodan et al. 1999). The Pasinler Basin occurs near the regional drainage divide. To the east, separated from the basin by a low col, are the headwaters of the Euphrates, which drains to the Persian Gulf, whereas catchments to the north drain into the Black Sea. The Basin itself is part of the headwaters of the River Aras, which drains east to the Caspian Sea.

The region occurs in the transition between true dry steppe to the east and more temperate climate zones to the west. The climate here is markedly seasonal, with continentality producing dry, hot summers and cold winters, often with heavy snowfall. This seasonality is reflected in surface hydrology, with river flows being highest in the spring and early summer whereas catchment soils are desiccated by late summer. Late-lying and permanent snow occurs at higher elevations.

In the past, a variety of climatic conditions have prevailed. During the Pliocene, conditions may have been broadly similar to that of the present interglacial, if perhaps a little warmer. During much of the Pleistocene, the region was significantly colder. Possible glacial landforms and sediments occur north of the basin, and the Pleistocene glaciation of Mt. Ararat, east of the basin, is well established (Altmli 1966; Birman 1968). There is little unambiguous evidence for glaciation within the basin itself, however.

2

Marine Ecosystem

Observational Evidence

We adopt the local stratigraphic nomenclature established by Terlemez et al. (1997). The oldest rock exposed along the Gölbasi-Türkoglu Fault is the Hatay Ophiolite, which was obducted onto the northern margin of the Arabian Platform during the Maastrichtian (c. 70 Ma), having apparently formed around 90 Ma at a spreading centre within the adjacent Southern Neotethys Ocean. Ophiolite obduction was followed unconformably by deposition of the Besni Formation, consisting of multi-coloured conglomerate and sandstone, passing upwards into Middle or Upper Maastrichtian limestone. This is followed by the uppermost Maastrichtian to Palaeocene Germav Formation, consisting of marl and clayey limestone, overlain by the Eocene to Upper Miocene Midyat Group carbonates. Along the Gölbasi-Türkoglu Fault, this is typically represented by the Lutetian Hoya Formation, a particularly pure, well-lithified limestone that resists erosion and forms much of the relief of the region.

Figure: *Marine Ecosystems*

Mid-Miocene uplift and sea-level fluctuation (e.g. Karig & Kozlu 1990; Arger et al. 2000) resulted in the interbedding of fluvial (Selmo Formation) and shallow marine sediments (e.g. Derman 1999). Contemporanous basaltic andesite volcanism (Arger et al. 2000). the 'Yavuzeli Basalt' (Terlemez et al. 1997), also occurred, resulting in a complex succession of lava flows interbedded with the sediments (e.g. Derman 1999). K-Ar dating by Arger et al. (2000) established the age of this volcanism as c. 16 Ma, consistent with the biostratigraphic age of the associated shallow marine sediments.

Before this dating study by Arger et al. (2000), this volcanism in the Pazarcik-Türkoglu area was widely thought to be Quaternary. However, the assumption of a Quaternary age was based on miscorrelation with other Quaternary volcanic fields elsewhere in SE Turkey (Terlemez et al. 1997). None the less, a Quaternary age continues to be claimed, and has been used to argue for a young (less than c. 2 Ma) age for the initiation of the East Anatolian Fault Zone. However, as dating shows that this volcanism is Miocene, such reasoning should no longer be used.

Likewise, some interpretations (e.g. those by Lyberis et al. (1992), Chorowicz et al. (1994) and Adiyaman & Chorowicz (2002)) have claimed, largely based on interpretation of satellite imagery without supporting field evidence, that the Gölbasi-Türkoglu Fault is a minor structure and the principal active deformation in its vicinity is shortening, leading to the development of the adjacent Ahir anticline and its counterparts. However, our field investigations, like those of Westaway & Arger (1996), reveal no evidence that these anticlines are active.

Their folding predates the East Anatolian Fault Zone, justifying the use by Westaway & Arger (1996) and in the present study of offset anticline axes as piercing points to quantify the young strike-slip. Our fieldwork also indicates that the interpretation by Adiyaman & Chorowicz (2002), that the Gölbasi-Türkoglu Fault is a normal fault with downthrow to the SSE, is incorrect, as no systematic component of vertical slip can be identified. In some places the topography across this fault is lower on its SSE side, but elsewhere the lower ground is on its NNW side; these local patterns can be readily explained by juxtaposition of different rock types, some more easily erodable than others, as a result of the active left-lateral slip.

The final stratigraphie unit, the 'Harabe Formation' of Terlemez et al. (1997), consists of fluvial gravel, sand and silt. This has been assigned a nominal 'Pliocene' age, as parts of it stratigraphically overlie the Yavuzeli Basalt. However, on the basis of this limited constraint, individual fluvial deposits grouped using this term could have any age from Late Miocene to Mid-Pleistocene. By analogy with the similarly designated 'Asartepe Formation' in western Turkey, it may well consist of a complex mixture of stacked fluvial units and inset deposits, forming high river terraces, such that it should arguably not be regarded as a 'Formation' in any meaningful stratigraphie sense.

Fault Zone Morphology

The Gölbasi-Türkoglu Fault can be readily identified for most of its length from the geomorphology. Its eastern end was described in detail by Westaway & Arger (1996). To the best of our knowledge, the only previous detailed investigation of the central and western Golbasi-Turkoglu Fault has been by Yalcin (1979). In the western Golbasi Basin the Golbasi-Turkoglu Fault strikes SW and runs for c. 10 km along the base of a c. 100 m high, NW-facing escarpment on its Arabian side, past Kosuklu, Catalagac and Kücükören. This escarpment exposes the uppermost Maastrichtian-Palaeocene clayey limestone of the Germav Formation; its base forms the alluvial plain in the basin interior at c. 880 m above sea level (a.s.l.), whereas its top truncates a low-relief land surface formed in the Germav Formation, which in places slightly exceeds 1000 m a.s.l.

The Kisik Drainage Catchment

About 1 km from the western end of the Gölbasi Basin, the River Aksu exits it through a gorge in this escarpment. At the entrance to this gorge (at [CB 6190 6725]) the Aksu is joined by the Kisik, one of its principal tributaries, draining an area of c. 40 km^2. From this confluence the Golbasi-Türkoglu Fault follows the Kisik upstream, passing between Sakarkaya and Soku, with a typical S60°W trend. The local relief is much more dramatic than farther NE, the escarpment on the Turkish side of the Gölbasi-Türkoglu Fault rising from c. 900 m to c. 1300 m a.s.l. as it truncates the axis of the Ahir anticline in Midyat Group limestone. West-away & Arger (1996) located this anticline axis at [CB 5950 6560] and regarded it as conjugate to the axis of the Körkün anticline on the Arabian side of the Gölbasi-Türkoglu Fault near the eastern end of the Gölbasi Basin.

The 33 km offset between these piercing points indicates the total slip on all Late Cenozoic left-lateral strike-slip faults that pass through the Gölbasi Basin, comprising the Gölbasi-Türkoglu Fault and the

NNE-SSW-striking fault zone that splays southward from it near the eastern end of this basin, which is not investigated in the present study. In contrast, the escarpment on the Arabian side of the Gölbasi-Türkoglu Fault maintains a roughly constant c. 120 m height, for instance rising from the c. 920 m river level to c. 1040 m a.s.l. where the Kisik River joins the fault line.

However, moving WSW, the exposure in the face of this escarpment passes down-section, from the Germav Formation into the underlying Besni Formation, recognizable from its distinctive coloration, and then into the Hatay Ophiolite. Tolun & Erentoz (1962) mapped the eastern end of outcrop of ophiolite on the Arabian side of the Golbasi-Türkoglu Fault in the vicinity of [PC 5640 6375], c. 1 km WSW of the point where the Kisik River reaches the fault line. East of this point, beyond the gorge flank on its Arabian side, outliers of the Besni Formation rise to c. 1100 m a.s.l.

The roughness of the local topography raises the possibility that these might overlie ophiolite, thus concealed beneath scree deposits, which may thus persist as far east as Y'. Furthermore, ophiolite is also present farther ENE in the escarpment face at least as far as [CB 5900 6550]. We return to the issue of how to determine precise piercing points in this ophiolite during discussion of the total slip on the Gölbasi-Türkoglu Fault. At the WSW end of the offset reach of the Kisik River, an offset reach of the C1narcik tributary follows the line of the Gölbasi-Türkoglu Fault for c. 2 km.

However, this is a minor drainage system compared with the Kisik trunk channel, draining only c. 3 km^2 of land area, and has evidently thus been unable to incise in pace with the regional uplift that has occurred. At its confluence with the Kisik it forms a hanging valley, where it cascades down to the level of the Kisik.

The Koca Drainage Catchment

Beyond the sector where the Cinarcik follows the line of the Gölbasi-Türkoglu Fault, this fault crosses the col, at c. 1240 m a.s.l., near Karaagac Tepe hill, entering the Koca river system. In this reach this fault typically trends S65°W, the outcrop being ophiolite on its Arabian side and partially lithified Miocene fluvial sand and gravel of the Selmo Formation on its Turkish side. Although the regional topographic gradient is southward, the more erosion-resistant ophiolite means that locally the land surface is typically higher in the immediate vicinity of the fault on its Arabian side.

The Gölbasi-Türkoglu Fault descends to the level of the Koca River and its principal tributary, the Sincer, along a gulley, passing directly south of Karaagac village and crossing the Velikler tributary, which is not significantly offset from the modern outlet gorge of the Sincer across the fault, reaching the Sincer at c. 905 m a.s.l. It then runs upstream along the Sincer, which is offset left-laterally by c. 800 m, forming a short section of linear valley between ophiolite and Selmo Formation sand and gravel.

Around the Golbasi-Türkoglu Fault leaves the Sincer River, passing for c. 1.2 km through a densely forested interfluve area, still forming the contact between ophiolite and Selmo Formation deposits before joining the Koca River at [CB 5270 6160]. Downstream of this point, the Koca flows within a gorge in the ophiolite on the Arabian side of this fault. This river is locally at c. 930 m a.s.l.; in the next c. 3 km upstream, through Kocadere, it follows the fault, rising to c. 980 m a.s.l. The land surface in the ophiolite on the Arabian side is locally at a relatively uniform altitude of c. 1070-1090m a.s.l. However, on the Turkish side, the landscape becomes progressively more subdued; at the downstream end of this reach the Selmo Formation is found at up to c. 1100 m a.s.l. whereas at its upstream end these deposits rise just a few tens of metres above river level.

Immediately WSW of the point where the Koca River joins it, at the Gölbasi-Türkoglu Fault enters a sector of very different relief, forming the contact between Midyat Group limestone on the Turkish side (Kardoga Tepe, rising to 1129 m a.s.l.), still with ophiolite on the Arabian side. The land surface in the ophiolite slopes gently westward from this point from c. 1070 m to c. 980 m a.s.l in c. 2.5 km distance, before the more deeply incised gorge system of the Gök River is reached; thus, the highest topography is now on the Turkish side of the fault.

The resulting col along the fault, at c. 1040 m a.s.l., is a distinctive landmark; dry valleys lead away from it both ENE towards the Koca and WSW towards the Gok. Westaway & Arger (1996) deduced that this localised Kardoga Tepe-Ahlankavagi Tepe outcrop of Midyat Group limestone. The c. 33 km offset between these piercing points confirms the total left-lateral slip through the Gölbasi Basin.

The Gok Drainage Catchment

The outlet of the Gök River from the Gölbasi-Türkoglu Fault, c. 1 km south of Kocalar, is a dramatic c. 120 m deep gorge, cut into ophiolite down to the c. 860 m a.s.l. river level. About 1 km farther east, truncated by the fault around, is an unnamed dry (or underfit) valley, c. 40 m deep, situated between the hills Tavsan Tepe at and Magara Tepe at, which leads into the Gök c. 1.5 km farther downstream. Two c. 1 km long gulleys join the fault line from the Midyat Group limestone of Ahlankavagi Tepe around, roughly in line with this dry valley, but with no surface drainage connection evident between the two. This Tavsan Tepe dry valley was evidently a former outlet for the drainage from the Turkish side of the fault, which presumably became abandoned as a result of the juxtaposition of the Kardoga Tepe-Ahlankavagi Tepe outcrop of Midyat Group limestone across it and the associated WSW offset of its former headwaters by slip on the Gölbasi-Türkoglu Fault.

For c. 10 km farther WSW, past Kocalar, Camhca and Sürülü, the Gölbasi-Türkoglu Fault (now oriented S57°W) is marked by the gorge of the Gök River, with ophiolite on the Arabian side and, initially, clayey limestone of the Germav Formation on the Turkish side. In this sector the Gök is joined by a succession of tributaries from the Turkish side, the Mezayok, Büyük/Sicanli, Cati, Cobanpinar, Karaagac and Alanyolu rivers. The farthest upstream of these, the Alanyolu, joins the Gök headwaters at, c. 8km WSW of the modern outlet gorge. The Germav Formation is more easily erodable than the ophiolite, so, immediately upstream of the c. 120 m deep Gök outlet gorge the higher topography is on the Arabian side of the Gölbasi-Türkoglu Fault.

However, farther WSW the land surfaces in the ophiolite and Germav Formation are both at similar levels, and the gorge depth remains a roughly uniform c. 120 m until the Karaagac confluence at, where this river is at c. 1030 m a.s.l. with the surrounding land surfaces at c. 1150m a.s.l. As in localities farther west on the Arabian side of the Gölbasi-Türkoglu Fault, the exposure in this Maastrichtian-Palaeocene sequence passes westward down-section, until ophiolitc is exposed beyond the Gök gorge to the north around.

However, the mapping by Tolun & Erentöz (1962) shows ophiolite persisting eastward within the northern flank of this gorge as far as. We postpone consideration of the potential value of these data as piercing points to constrain the total slip on the Gölbasi-Türkoglu Fault to later discussion.

The western Golbasi-Türkoglu Fault. For c. 10 km beyond the western margin of the Gök catchment, the Gölbasi-Türkoglu Fault is confined within the Hatay ophiolite, with a typical local trend of

S78°W. There is little fluvial incision near the fault in this sector, and no sites where significant river offsets are clearly defined, evidently as a result of the erosion resistance of the ophiolite and the limited erosional power of these headwaters. Both the fault-line valley and the surrounding landscape reach their highest altitudes in this sector, possibly because the ophiolite is locally at its thickest.

The headwaters of the Gök drain ENE from, leaving the eastern part of an expanse of linear valley floor, Göl Alani, situated at c. 1250 m a.s.l., which forms the col between the Gök catchment and that of the Gökgecit River farther WSW. In this sector, the surrounding land surface in ophiolite reaches c. 1310 m on the Turkish side of the fault, with many hilltops at similar altitudes on the Arabian side, including Catalbükü Sirtlan, Gökgecitbasi Tepe, and an unnamed c. 1300 m summit south of Göl Alam at [CB 3845 5355].

The western end of Göl Alam is drained WSW from by the Gökgecit River, which follows the fault for c. 4 km to Kartal, leaving the fault line on its Arabian side at [CB 3415 5320]. Kartal village is located on the sloping surface of an alluvial fan on the Turkish side of the fault, which reaches as low as c. 1050 m a.s.l. in the vicinity of the Gökgecit outlet gorge, where this river has incised to c. 100 m lower. Just west of Kartal, at [CB 3405 5320], the Gölbasi-Türkoglu Fault crosses a low col in the alluvial fan surface, which forms the drainage divide between the Gökgecit and Cigli rivers. From this point, the Cigli River follows the linear valley along the fault to Türkoglu.

Around [CB 3115 5165], c. 1.5 km ENE of Cigli (or Ayanusagi) village, the fault ceases to be confined within ophiolite on its Arabian side. For c. 2 km from this point it passes through an area mapped by Terlemez el al. (1997) as an outcrop of the 'Harabe Formation', but which seems to be more appropriately described as another large alluvial fan, shed from the ophiolite to the north, through which protrude inliers of ophiolite and Besni Formation sediment. This locality, where ophiolite ceases to be juxtaposed on the Arabian side of the fault, was used by Westaway & Arger (1996) as a piercing point to infer 16 km of total slip, relative to the western end of the ophiolite outcrop north of the fault at Türkoglu. However, as with other possible piercing points related to margins of ophiolite outcrops, the precise location of this one is difficult to establish: it could arguably be anywhere from [CB 3115 5165] (noted above) to [CB 2950 5100], representing a projection onto the fault of the westernmost outcrop of ophiolite, in the vicinity of Cigli village. Beyond Cigli, the Gölbasi-Türkoglu Fault can be traced near the foot of the ophiolite escarpment on its Turkish side, against the Quaternary alluvial plain of the Cigli River. At [CB 2680 5080], c. 3 km WSW of Cigli, where transected by the road from Kahraman Maras to Gaziantep, this plain is c. 600 m a.s.l. and the adjacent ophiolite rises to c. 700 m a.s.l. Moving WSW, the surface to the plain descends gradually to c. 470 m a.s.l. at [CB 1450 4250] near Türkoglu, where the Aksu River crosses the fault from south to north. In contrast, the ophiolite rises to 903 m a.s.l. at the summit of Koroglu Tepe [CB 1990 4975] before gradually descending to beneath the Quaternary alluvial plain around Turkoglu.

The Gölbasi-Turkoglu Fault can be clearly traced near the base of this escarpment past Tevekkelli, Kocalar and Öksüzlü, with a typical trend of S62°W, a clear fault scarp being evident for c. 2 km around Tevekkelli between [CB 2515 4950] and [CB 2360 4870]. Pervasive cover by Pleistocene slope deposits conceals the ophiolite west of c. [CB 1770 4550], near Turkoglu. However, directly downstream of the line of this fault the Aksu at c. 470 m a.s.l. is flanked by bluffs rising to c. 540-550 m a.s.l., around [CB 1550 4450] in its right bank and as far west as [CB 1300 4300] in its left bank, which may represent the westernmost part of this ophiolite unit, obscured beneath Pleistocene deposits. As with other sites already discussed, use of this margin of the ophiolite as a piercing point for estimating the total slip on the Golbasi-Turkoglu Fault is thus somewhat problematical.

Late Holocene Landscape Evolution

Archaeological excavations at the coast of A'asu, in Tutuila Island of American Samoa, exposed a depositional sequence spanning the past circa 700 years. With the period represented, sedimentation rates exceeded 10.15 cm per century in the valley floor and 16.34 cm per century along the valley margin. The occupational history may correlate with changes in climate, sea level, and coastal geomorphology. Although the evidence accords with the expected responses to the Little Climatic Optimum (circa 1050 to 690 B.P.) and Little Ice Age (circa 575 to 150 B.P.), the most plausible explanation for the A'asu case is that environmental change accompanied expansion of upland land use. Based on evidence here and elsewhere in Tutuila, it is proposed that the establishment of fortifications, monuments and permanent settlements in the uplands was part of a broader pattern of land-use expansion beginning in the fourteenth century A.D. Thirty-five years ago, a study of oxygen-isotope ratios from a speleothem (cave formation) in New Zealand demonstrated that a rapid temperature drop occurred in southern Polynesia in the fourteenth century.

Wilson et al. (1979) proposed that the violent climatic conditions accompanying this temperature drop would not only have resulted in increased latitudinal temperature variants but would also have had catastrophic effects on agriculture. At the same time, the more violent climate would have made long-distance voyages inherently more dangerous. More recently, Bridgeman (1983), based on an extensive review of previously published climatological data, proposed that climatic change may have contributed to a general collapse in Polynesian migrations after A.D. 1350. These studies have received renewed discussion since 1995 when Nunn (1995, 1998, 1999, 2000) published sea level data synchronised to climate change, proposing his own theory that changing climatic conditions had a profound regional effect on Polynesian culture.

A better understanding of climate-induced landscape change is necessary in order to better model human response to dynamic ecosystems. Climate change may result in systematic environmental adjustments in an alluvial catchment. Because archaeological sites are found within Holocene-age sediments, it is necessary to understand the late Quaternary history of the region in order to understand prehistoric settlement patterns and the completeness of the archaeological record.

In order to determine the extent to which changes in the late Holocene geological record correlate with climate and ecological changes over the same period, geoarchaeological explorations were undertaken in A'asu on the remote northern shores of Tutuila Island, American Samoa. As the geochronology became known, it was apparent that the sedimentary record had been highly irregular in the late Holocene, particularly after A.D. 1300. Explaining the causes of that dynamism became the focus of research over the next several years. In this chapter, geoarchaeological data from A'asu are compared with similar data from coastal settings and combined with recent research on the establishment of mountain settlements Pearl 2004) to show that changes in climate, settlement patterns, and landscape evolution converged, beginning in the fourteenth century A.D., a time when social complexity was on the rise in Samoa. The lush valley of A'asu is situated on the north coast of western Tutuila, with numerous inhabitable valleys undisturbed by roads and contemporary settlements, standing in stark contrast to other more populated and developed areas of Tutuila. The valley floor extends about 0.5 km inland and comprises about 0.1 [km^2] of mostly gently sloping terrain. A considerable freshwater stream, derived from runo off and several springs high in the mountains, dissects the valley floor. The springs provide a perennial source of fresh water and likely made this valley particularly attractive to prehistoric inhabitants.

A'asu provided a fortuitous starting point for geoarchaeological study because its constricted valley opening yet fairly large catchment cause it to act like a "graduated cylinder," accentuating periods of erosion and stability in the stratigraphic record. The stratigraphy of A'asu Valley is seen as a component piece in the larger picture of late Holocene landscape evolution in eastern Samoa.

The valley is formed of alluvial sediments transported along A'asu Stream and its tributaries and colluvium eroded from the steep-sided valley walls. Slope-eroded colluvium has created thick deposits along the valley floors and margins that interfinger with alluvium deposited by the stream. Along the coastal plain, sediments have been reworked by wave action, storms, and fluctuations of sea level.

A village at the mouth of the valley was encountered in 1787 when the French explorer La Perouse made landfall on Tutuila. At their meeting in 1787, Samoans and French explorers clashed at A'asu, resulting in the deaths of an undetermined number of Samoans, the French expedition's second in command (de Langle), and 11 other members of the landing party. Today there are no permanent inhabitants of the village, but it is a popular site for seasonal fishing activities. Taro and other cultigens are farmed nearby.

Emerged reef fragments visible from A'asu and at more than 20 other locations around the island indicate that relative sea levels are lower at present than earlier in the Holocene. Stearns (1944) was the first to consider that these were evidence for a higher sea stand, but their late Holocene dates were not known until recently. Nunn (1998) dated six reef fragments ranging from 0.75 to 2.11 m above mean sea level, each of which dated to the last 1000 years.

Similar data have been obtained in Western Samoa, where raised beach rock indicates a mid-Holocene highstand of 0.8-2.3 m at 1200-1850 [+ or -] 70 years B.P. Independently, Dickinson (2001) reports the differential elevation between modern and emergent wavecut coastal benches offshore of Tutuila (on the islet of Aunu'u) at 1.8 [+ or -] .01 m, a figure very close to the calculated theoretical value of 1.9 m. Furthermore, coastal progradation on 'Upolu since 700-1000 years B.P. has been attributed to systemic adjustments caused by lowering sea levels. This interpretation underscores the need for additional studies on the impact of climate and sea level change on landscape evolution in Samoa. Though some islands in the Samoan chain are experiencing subsidence or upflexure due to volcanic activity, the islands of Tutuila and Aunu'u are not thought to have experienced any significant vertical change.

Eustatic shoreline movements on Tutuila are key to interpreting the geomorphic changes at A'asu. Comparing an elevated beach rock geochronology with data from 18 other tectonically stable sites in the Pacific, Nunn (1998) summarised eustatic shoreline movements on Tutuila during the last millennium as follows: (a) a sea level rise, coincident with a period of warming known as the Little Climatic Optimum, between 1050 and 690 B.P.; (b) a sea level drop during the Little Ice Age, between 575 and 150 B.P.; and (c) a period of recent warming during the last 150 years. The transition between Little Climatic Optimum and Little Ice Age, which peaked between 690 and 575 B.P., was marked by rapid cooling and possibly a concurrent increase in rainfall.

A study of the geochronology of A'asu was conducted simultaneously with archaeological investigations; consequently, geological profiles discussed herein are those of the archaeological excavation units (sondages). The work was completed in two eight-week field seasons in the months of May and June of 2001 and 2002. The research was initiated by the author as the first step in a multiyear research plan to better understand the late precontact archaeology of Samoa. The fieldwork was conducted by the author, several graduate and undergraduate students, and professional excavators. Because the final report of the archaeological investigations is still in preparation, the general methods and procedures are summarised below.

The archaeological plan called for multiple 1 x 1 m sondages, sometimes contiguously grouped into blocks, to be located across the landscape. Fifteen such sondages were tested in this manner and labelled blocks A through E. Of those blocks, this study will focus only on blocks A and D, which provided deep exposures of alluvium exhibiting stratification and abundant charcoal for dating (the remaining blocks focused principally on surface deposits). Block A was situated inland near the valley margin in a setting that was expected to have abundant colluvium, while block D was situated on the modern floodplain of A'asu Stream. Both of the blocks were located in depositional settings, but it was anticipated that block A would reveal higher sedimentation rates due to its location at the toe of a steep valley wall. Ultimately, block A was expanded to include two adjacent sondages, reaching a maximum depth of 2.0 m. Block D was expanded to include five contiguous sondages, reaching a maximum depth of 1.3 m. Excavations proceeded in arbitrary 10-cm levels or by natural stratigraphy when change occurred within these. All sediments from block A (excavated in year 1) were screened through 1/8" mesh, while 1/4" mesh was used in block D (excavated in year 2). The change was made between seasons after careful consideration of the materials being recovered and time available to us in the second season. Water screening was employed at block A when the clay fraction exceeded about 30 percent. Artifacts were point-plotted before removal when possible. A stratigraphic cross-section was made for each completed excavation block, which was then photographed and backfilled.

Sediment samples were routinely taken at least once every 10 cm. Charcoal was so ubiquitous that no fewer than four samples were taken every 10 cm. The process of washing, sorting, cataloging, and analysing artifacts began in the field and was completed in the archaeology laboratory at Texas A&M-Galveston. Soils and sediments exposed in sections were described using standard procedures and terminology outlined by Soil Survey Staff (2003). Major lithostratigraphic distinctions are made by grouping substrata into major strata, numbered consecutively I, II, and III. Subtle changes within these strata are given an alphanumeric designation. For interpretive purposes, the conventional radiocarbon age was calibrated with the OxCal 3.10 radiocarbon calibration software using the INTCAL04 atmospheric carbon curve for calibration.

Stratigraphic Sequence

Block A

The two sondages comprising block A were situated directly on a stone-lined house platform. Such platforms were the predominant foundation type in ethnohistoric accounts of Samoa, providing foundations for dwellings, ceremonial guest houses, cooking houses, and even churches and other buildings. This one, like most others, is covered in a pavement of gravels ('ili'ili) and bordered by one course of basalt stones. Most of the sediments encountered were terrigenous. However, some sands, corals, and shells were intermixed with the alluvium and colluvium, especially in the uppermost strata. Soil horizons (pedostratigraphy) and lithological discontinuities (lithostratigraphy) were both noted. Based on their lithology and depositional features, ten divisions were recognised within the three major strata. Strata Ia-Ie—Strata Ia through Ie re.ect the construction sequence of the house platform, with Ie consisting of loose coral and basalt cobbles representing the .rst stage of construction. The contact between strata Ie and IIa beneath is abrupt and irregular, indicating that some erosion or intentional clearing may have preceded the deposition of Ie.

A conventional radiocarbon determination of 100 [+ or -] 1.4 pMC (A-12406) was obtained on charcoal from the bottom of stratum Ie. Taken at face value, 100 percent modern carbon would indicate a date of about A.D. 1950. However, the lower limit (2-sigma) calculates to the interval A.D. 1660-present. The upper limit falls in the mid-1950s if all the carbon was fixed in a single year. Better precision is not possible in this interval of time because of high atmospheric carbon variations. The corrected age for this sample is equivalent to 0 [+ or -] 110 B.P. Most likely the structure dates to A.D. 1800-1930, although an earlier date cannot be ruled out.

Strata IIa-IIb—Stratum II consists predominately of clay loams, but a significant fraction of pebbles, gravels, and shells make the group quite heterogeneous. Stratum IIa is made of very dark grayish-brown sandy clay loam, with as much as 25 percent of the matrix pebble and larger-sized clasts. Most pebbles are subrounded, indicating not only their colluvial origin but that they have rolled some distance prior to deposition. There are abundant marine shells throughout these sediments, many exhibiting burning, and high densities of other cultural artifacts, though their numbers decrease with depth. Abundant charcoal and bone were also recovered at the contact with the overlying stratum. Stratum IIb transitions smoothly over a 10-cm boundary. This is a pedogenic distinction; that is, the boundary was formed through natural soil-forming processes. It is identical in colour and texture to IIa and is distinguished principally by soil structure and the rapid reduction in the number of marine shells, though they still constitute a small fraction of the horizon.

A radiocarbon age of 355 [+ or -] 35 B.P. (AA-51255) indicates that deposition of stratum II was under way in the fifteenth century. It is separated from stratum III by an abrupt irregular boundary, an indication that erosion occurred between the two. Strata IIIa-IIIb—Stratum III consists of fairly homogenous sandy clays, with clay content slightly increasing with depth. The principle differentiation between sub-horizons a and b is that the latter is slightly darker (dark brown vs. brown) and has higher clay content (sandy clay vs. sandy clay loam). The changes in colour and texture are related and occur gradually with depth. Stratum III is generally more fine grained than the overlying strata, suggesting, in the absence of any evidence for pedogenically altered clays, that water transport played a greater role in depositing the sediments. However, occasional angular rocks and a large boulder encountered at the base of block A reveal that much of the deposition is still colluvial (from the slopes) rather than fluvial (from the flooding of A'asu Stream). Stratum IIIb is the deepest and oldest encountered, and there is abundant charcoal dispersed in this layer.

Four radiocarbon ages were determined on charcoal from stratum III. Two samples were taken from the burned hearth mentioned above. A third and fourth sample were taken from dispersed charcoal beneath the hearth area elsewhere in the layer. The samples were separated by a distance of no more than 20 cm. The two samples from the hearth returned calibrated radiocarbon ages of 625 [+ or -] 35 B.P. (AA-51257) and 630 [+ or -] 40 B.P. (BETA-171844). The samples beneath the hearth returned calibrated radiocarbon ages of 635 [+ or -] 35 B.P. (AA-51256) and 710 [+ or -] 40 B.P. (BETA-171845).

Block D

The five sondages comprising block D were situated on the alluvial plain. At first glance it appears to be an area devoid of architectural features, and initially the first sondage was placed there to establish the underlying stratigraphy of the alluvial plain. However, archaeological discoveries necessitated expanding the block to a full 5 [m^2]. Cultural materials were encountered in each stratum, including an adult human burial in IIIa. Based on their lithology and depositional features, seven divisions were recognised within the three major strata. Like block A, most of the sediments encountered were terrigenous. However, some marine sands, corals, and shells were intermixed with the alluvium and colluvium, especially in the uppermost strata, such as Ia below.

Stratum Ia-Ic—A 5-cm veneer of sand coats the surface. This sand, comprising stratum Ia, likely resulted from hurricanes that blasted this coastline in the 1990s. Former residents report that the village was leveled at that time and significant volumes of seawater flooded the low-lying parts of the village.

Stratum Ia abruptly overlies stratum Ib, a very dark grayish-brown sandy loam. Pebbles and gravels make up to 20 percent of the matrix, which also contains small amounts of fire-cracked rock and charcoal.

Stratum Ic represents a single anthropogenic horizon. Its coral cobbles and matrix of gravelly, sandy clay loam was initially encountered in the floor of some sondages. As the excavation area was laterally expanded, it became clear that this clast-supported matrix was bounded on the south by a line of basalt boulders. These were 20-30 cm across and are consistent with those seen bordering domestic areas in ethnohistoric contexts. The floor was punctuated by a cobble-lined pit feature extending into the underlying strata.

Strata IIa-IIb - Directly beneath the contact of stratum I, a soil-enriched, very dark grayish-brown sandy to gravelly clay loam (up to 20 percent pebbles and gravels) was encountered (stratum IIa). Many marine shells were also encountered, as well as some bone, charcoal, and other cultural material. This stratum, distinctly darker than the overlying layer, could be seen clearly following the outline of pit features in adjacent sondages. Charcoal from the base of IIa provided a radiocarbon age of 320G35 B.P.

Stratum IIb transitions smoothly over a 10-cm boundary. It is distinguished from IIIa by a slightly lighter value (dark grayish-brown), as well as by decreasing gravels, shells, and cultural finds. It terminates abruptly and lacks any soil development. Charcoal from the base of this stratum provided a radiocarbon age of 340G50 B.P.

Strata IIIa-IIIb—Stratum III consists of homogenous sandy clays, with clay content slightly increasing with depth. It is lithologically distinctive from stratum II; the abrupt planar boundary between them represents an erosional disconformity. The principle differentiation between IIIa and IIIb is that the former is strongly discolored by abundant charcoal. The entire horizon can be characterised as an ashy mass of poorly sorted, decomposing charcoal fragments in sandy clay, with some lithics and faunal remains. It would not be appropriate to refer to the layer as a lens because its lateral extent remains unknown. Further, there was no evidence of any depression or depositional cavity that might have bounded the layer. The lower boundary was clear but not abrupt, with a transition to IIIb visible over about 5 cm. Charcoal content in the underlying stratum was dramatically lower.

Stratum III is generally more fine grained than the overlying strata, suggesting that fluvial transport played a greater role in depositing the sediments. Stratum IIIb is the deepest and oldest of the sedimentary units. It contained abundant cultural material as well, including an adult burial, a fishing weight (not associated with the burial), and other cultural materials. A single radiocarbon age was determined on charcoal from stratum III. A large sample of concentrated charcoal from IIIa returned an AMS age of 650G50 B.P.

Interpreted Geochronology of A'ASU

Correlation of the stratigraphy between blocks A and D is based on the lithology and soil morphology, and major strata given the same designation (i.e., I, II, or III) are interpreted as being part of the same depositional layer. Atmospheric radiocarbon fluctuations between A.D. 1325 and 1375 make calibration of dates in the fourteenth century problematic. Conventional radiocarbon ages between about 500 and 700 B.P. have a bimodal probability distribution when calibrated. Furthermore, many of the dates that I cite were calibrated using different calibration curves than are now available. Consequently, for comparative purposes I will emphasise the ages in radiocarbon years (B.P.), which are consistently reported in the papers cited and are independent of calibration curves.

These dates essentially cluster into three groups. A chi-square of the five dates from stratum III reveals that they are statistically the same at the 95 percent confidence interval (648 [+ or -] 17; df = 4; T = 3.2; [chi square] (.05) = 9.5). Similarly, a chi-square of the three dates from stratum II reveals that they are statistically identical at the 95 percent confidence interval (341 [+ or -] 24; df = 2; T = 0.4; [chi square] (.05) =4 6.0). This gives a high degree of confidence that the gap between stratum II and III is real and suggests that the vacuity represents about 300 years.

The oldest radiocarbon date from A'asu is 710 [+ or -] 40 years B.P., obtained from charcoal in stratum IIIb in block A. A small number of lithics, including an adze fragment, were recovered beneath this charcoal sample. Utilisation of A'asu Valley therefore began prior to 710G40 B.P. During the Little Climatic Optimum (~1050 B.P. to ~690 B.P.), the predominance of well-sorted, fine-grained alluvium in the valley suggests that fluvial processes predominated. The dispersed charcoal in the deepest sondages probably indicates that swidden agriculture played an important role in subsistence patterns at that time, though .re from natural causes cannot be ruled out. It should be noted, however, that while human use of fire for agricultural purposes in moist tropical environments is very common, large natural fires are very uncommon.

Around 650 [+ or -] 50 B.P., the valley floor was blanketed by charcoal, at least locally, after which the surface was truncated by erosion. The precise timing of this erosional event is unknown because the sediments related to it have been flushed out of the valley. The direct source of the charcoal is also undetermined, but erosion across the valley floor may be due to decreased vegetation cover. A date on the overlying stratum of 355 [+ or -] 55 B.P. means that the lacuna represented by the erosion includes the transitional period between the Little Climatic Optimum and the Little Ice Age of the Late Holocene (~690 to ~575 B.P.) and might be directly correlated to it.

Deposition of the stratum II was under way no later than 355 [+ or -] 35 B.P. On the valley margin, rapid colluvial deposition predominated, with minimal alluvium, while deposition on the valley floor was a mixture of colluvium and alluvium. The increased sediment yield was coincident with the Little Ice Age, when both temperature and sea level were lower than present. A mean rate of sedimentary deposition was calculated by dividing minimum vertical accretion (depth between the bottom of stratum I and the deepest radiocarbon determination in each block) by mean accumulation time (as determined by the radiocarbon intercept).

Due to the contribution from the steep valley walls, rates of sedimentary deposition along the valley margins were 62 percent higher than on the valley floor. Along the valley margins, sediments accumulated at a mean rate of 16.34 cm per century (116 cm over 710 years). Of course, because of the lacuna left by erosion, the actual rate of deposition was higher at times. Along the floor of the valley, sediments accumulated at a mean rate of 10.15 cm per century (66 cm over 650 years). While perhaps over-generalised, these calculations indicate that deposition has been significant over the last 700 years.

Throughout the Holocene, sea level constrained the movement of sediment towards the sea. During the Little Climatic Optimum, the alluvial plain aggraded. Then, as the climate transitioned to the Little Ice Age, A'asu Stream cut into its alluvial plain and its gradient increased as it adjusted to a lower sea level. Sea level rise over the past 150 years or so has begun to stabilise the alluvial plain. In Tutuila and Aunu'u, the drop in sea level after the mid-Holocene high-stand exposed more landmass and created conditions for prograding shorelines.

The aceramic site of A'asu has a rich historic and prehistoric archaeological record. So that readers may better understand the archaeological context, a brief summary of the archaeology is provided here. The modern surface is a palimpsest of recent and ethnohistoric discard. Secondary maintenance has removed most artifacts from major walkways, but discarded materials are found around many of the house foundations and at the periphery of the village. Discards range from wire, glass, and metal "trash" to the occasional polished stone flake or adze fragment. Most of the existing house foundations are cement slabs of recent origin, though there are several traditional stone fale foundations as well.

The archaeology of the uppermost excavation layers (stratum I) is closely related to recent village activities. Modern and historic glass, metals, and plastics were recovered, including such diagnostic objects as nails, buttons, and religious devotional medals. These artifacts are a clear indication that the site was well connected with long-distance trade networks in historic times. Many of the artifacts are marine related-boat-building nails (copper with roves), shell-motif beads, and shell buttons-attesting to the strong connection that the people of A'asu had with the sea.

The archaeology of stratum II was increasingly prehistoric in character, consisting principally of lithics and faunal remains that declined with depth, an indication of the poor preservation conditions at the site. The lithic artifactual remains, however, showed a bias towards non-adze stone tools. A number of worked flakes, especially unifacial scrapers, were encountered, as well as a small number of adzes. Use-wear indicates that the adzes were usable finished products, but they were not highly polished like many other Samoan adzes. Stratum II contained a limited number of historic artifacts, including a kaolin clay pipe fragment of a style that was common at the end of the eighteenth century.

The exclusively prehistoric archaeology of stratum III was very similar to the archaeology of stratum II. Indeed, as of this publication I am not prepared to say that there was a significant difference in terms of lithic technology, and even fewer faunal remains were present. An oval stone "weight" was recovered that might have been a fishing or net weight. A burial was also recorded in block D, stratum IIIb. No cultural materials were associated with the burial, which was lying on its right side—facing north, towards the sea—in a flexed position.

Cultural Stratigraphy and Site Formation Processes

Three archaeological components are present at Masterov Kliuch. Component I is an early Upper Palaeolithic occupation occurring from

90-100 cm below the surface within geologic unit 2. Component II also appears to be an early Upper Palaeolithic occupation; it occurs within unit 4 at a depth of 30-60 cm below the surface. Component III is a Bronze Age occupation ranging in depth from 0-20 cm below the surface within units 5 and 6. Given the complex geologic context of the Masterov Kliuch site, an important part of our research has been to establish the integrity of the site's Palaeolithic components, especially in terms of natural site deformation processes related to colluviation and cryoturbation.

Three indicators of site integrity—vertical distribution of artifacts, horizontal distribution of artifacts, and presence of conjoined artifacts—were studied in order to ascertain the degree of disturbance by colluvial processes. For components I and III, vertical distribution of artifacts is relatively tight, with component I occurring within a 10-cm-thick band and component III occurring within a 12-cm-thick band. A similar pattern can be seen in the horizontal distribution of artifacts from components I and III, with artifacts being situated in identifiable clusters across the excavation. Further, 13 artifacts from component I were conjoined; the average horizontal distance between these conjoined artifacts is 20.25 cm, and the average vertical distance is only 1 cm. The tight vertical and horizontal distributions of artifacts, as well as the close horizontal and vertical proximity of conjoined pieces, suggest that the artifacts of component I lie in a primary context.

Component II artifacts, however, have a much greater vertical distribution than those in components I and III (component II has an average thickness of about 20 cm), and the horizontal distribution of artifacts appears more scattered than in components I and III. Further, no artifacts from component II could be conjoined. These data suggest that component II is redeposited. Frost-heaving (the movement of artifacts due to repeated freezing and thawing of sediment) is also a factor affecting northern archaeological sites. To evaluate the degree to which freeze-thaw processes impacted the cultural components at Masterov Kliuch, we measured trend and plunge of all large-sized artifacts encountered in situ in the Palaeolithic components, using a Brunton pocket transit. Twenty-two such artifacts were analysed in this way—16 for component I and 5 for component II.

Plunge measurements show that roughly half of the artifacts lie within 45 [degrees] of horizontal; that is, they lie more flat than upright. The other half lie more vertically upright, with plunge measures of between 45 [degrees] and 90 [degrees]. Of these, only three artifacts have plunge measurements of 90 [degrees]. Once reaching 90 [degrees] plunge, artifacts tend to move upward through the profile. Thus, although frozen ground processes appear to have reoriented some artifacts, there is little indication that they have displaced them vertically. Trend measurements, further, show no obvious pattern in the direction that the artifacts plunge, and few actually are trending along the slope of the site (about 100 [degrees] east of north), suggesting that artifacts of component I have been reoriented by minimal frost-action, but probably not slumping or slopewash.

The stratigraphic and provenience information from Masterov Kliuch show that while both Palaeolithic components lie in colluvial deposits, only component II is redeposited. Component I appears to lie in its primary place of deposition. Although frost-heaving has affected the orientations of the artifacts from components I and II, this process does not appear to have affected the locations of these artifacts.

Radiocarbon Chronology and Age of Cultural Components

Samples of bone (n = 3), tooth enamel (n = 1), charcoal (n = 1), and soil organics (n = 3) from the geological units and cultural components at Masterov Kliuch were dated through accelerator radiocarbon (AMS^{14}C) procedures. Charcoal was not well preserved in the site, occurring only in the uppermost unit in association with archaeological component III. For this reason, we concentrated on the dating of bone and other materials. When appropriately purified using XAD-2 resin, bone with significant amounts of intact collagen (typically greater than 5 percent of original amount of protein) can provide accurate age estimates (Taylor 1997). Pretreatment and AMS ^{14}C analysis of all samples was conducted at the NSF-Arizona AMS Facility, following standard methods described by Long et al. (1989) and Jull et al. (1983) for the AMS ^{14}C dating of bone and charcoal, respectively. All dates are reported as uncalibrated.

Five radiocarbon ages were obtained from stratigraphic unit 2. Three samples of bone, one sample of tooth enamel, and one sample of organic matter were dated. The most reliable ages were derived on the three bone samples. Two of the bones were collected from archaeological component I during our excavations in 1996. These XAD-purified samples of bone (AA-23640 and AA-23641) had relatively high amounts of original protein (11.3 and 14.8 percent, respectively) and yielded ages of 32,510 [+ or -] 1440 and 29,860 [+ or -] 1000 B.P., which overlap at two-sigma. The other XAD-purified bone sample

(AA-8888), which yielded an age of 24,360 [+ or -] 270 B.P., was collected from Meshscherin's 1991 excavation (Goebel 1993). This sample was collected from stratigraphic unit 2, but above archaeological component I and may be from a later brief occupation. These dates clearly indicate a pre-Sartan (pre-late glacial) age for unit 2 and that archaeological component I dates to roughly 30,000 B.P.

A pre-Sartan age for unit 2 is also supported by the frost cracks that extend from the top of this unit as well as the absence of frost cracking in overlying units. These frost cracks probably developed during the Sartan glacial period, as at other Upper Palaeolithic sites in the Baikal region (Bazarov et al. 1982; Tseitlin 1979). Thus, unit 2 and its associated archaeological component must pre-date the Sartan glacial period based on geologic evidence. This geologic scenario is supported by a radiocarbon age on organic-rich sands found within a channel overlying the frost-cracked surface of unit 2 in a test pit 50 m to the northwest of the main 1996 excavation area. Organic matter from this sand yielded an age of 18,850 [+ or -] 135 B.P. (AA-23647). This organic material appears to have been derived from the erosion of soils that had developed on the slopes above the site.

Two aberrant ages were obtained from stratigraphic unit 2. One small fragment of tooth enamel from archaeological component I yielded an age of 19,415 [+ or -] 260 B.P. (AA-23642). Also, a small fragment of what was thought to be charcoal was collected near a frost crack in the sidewall exposure of Meshcherin's excavation. This sample yielded an age of 18,335 [+ or -] 320 B.P. (AA-23643). These ages are at odds with the older bone-derived ages from component I and the geologic evidence. There are several reasons why these younger ages are disregarded. The age of 19,415 [+ or -] 260 B.P. (AA-23642) was derived on the inorganic apatite fraction of the tooth. Apatite is notorious for yielding inconsistent results because a number of mechanisms can significantly alter carbon-isotope values in the apatite structure.

The date of 18,335 [+ or -] 320 B.P. (AA-23643) turned out not to be derived from charcoal, but instead from an aggregate of organic matter. We believe this aggregate most likely represents the post-depositional movement of an organic particle into unit 2. Since this sample was collected only 2 cm from a visible crack, it may have been translocated into unit 2 from higher in the profile. As mentioned above, in some places on the site organic-rich sands dating to 18,850 B.P. are found overlying unit 2. Both the date on organic matter from unit 2 and the date from the overlying organic-rich sand (in the nearby test pit) are statistically indistinguishable at one-sigma. It seems likely that a sample of this organic-rich sand was translocated downward through the profile via a frost crack into the underlying unit 2. Thus, this age is considered invalid.

In an attempt to date geological unit 1, at the base of the profile, a bulk sample of soil organics from geologic unit 1 (taken from about 150 cm below surface) was AMS [sup.14]C dated and yielded an age of 7630 [+ or -] 65 B.P. (AA-23646). This date is clearly too young based on the overlying dates from unit 2 and can be disregarded. Stratigraphic units 3, 4, and 5, and archaeological component II are undated. Based on the organic-rich sand age and artifacts from component II, these appear to date to the late Upper Pleistocene, perhaps 18,000-10,000 B.P.

A sample of charcoal from a small hearth feature in component III near the top of the stratigraphic profile yielded an AMS [sup.14]C age of 2895 [+ or -] 45 B.P. (AA-23648), providing support for the presumed late Holocene age of this cultural component. Given these AMS [sup.14]C determinations, as well as the above review of site stratigraphy and site formation processes, we can make the following conclusions about the age of the Masterov Kliuch sediments and cultural components. Unit 1 was deposited sometime prior to 30,000 B.P. The frost cracks and small ice-wedge pseudomorphs that originate along the upper contact of unit 1 perhaps formed during the Konoshchel'e cold snap, dated elsewhere to 33,000-31,000 B.P., or during some earlier stade of the early or middle Pleniglacial. Unit 2 and component I are AMS ^{14}C dated to about 30,000 B.P., the beginning of the Lipovsko-Novoselovo interstade (independently dated to 30,000-22,000 B.P.). This is further supported by the extensive network of frost cracks and ice-wedge pseudomorphs that originate from the upper contact of unit 2; these probably formed during the height of the Sartan stade (22,000-17,000 B.P.). Unit 3 and component II have not been AMS ^{14}C dated, but, given their stratigraphic position above features relatively assigned to the last glacial maximum, as well as the platy structure of the sediment, must have been deposited (from upslope) sometime during the late glacial (17,000-10,000 B.P.). Units 4, 5, and 6 likely formed during the Holocene (10,000 B.P. to the present). Component III, found within units 5 and 6, dates to about 3000 B.P. and thus can be assigned to the Transbaikal Bronze Age.

Archaeological Assemblages and Features

The two Masterov Kliuch Palaeolithic components are described in detail below. Because it was not a focus of our study, the Bronze Age component is only briefly presented. For definitions of terms used to describe cores, tools, and other lithic artifacts, readers are referred to Andrefsky (1998) and Goebel (1993).

Component I

Cultural component I consists of a relatively dense band of lithic artifacts, with two distinct concentrations occurring in the 6-[m.sup.2] excavation, including a small cluster of flaking debris in the northwestern corner of the excavation (square 26T), and a larger cluster of retouched artifacts, cores, and flaking debris in the eastern half of the excavation (squares 25P, 26P, 25R, 26R). Within the latter cluster, two lithic technological activities are evident: (1) primary reduction activities represented by a concentration of 6 cores and about 60 cortical flakes, and (2) secondary reduction activities and tool use represented by a concentration of 18 retouched artifacts and nearly 30 retouch chips.

The few fragments of bone that were encountered during excavation of component I came from the eastern half of the excavation, in association with the concentrations of tools and retouch chips. The component I assemblage consists of 367 pieces, including 360 lithic artifacts and 7 small bone fragments. Among lithic raw materials, dark gray cryptocrystalline silicate (ccs) dominates, making up 73 percent of the assemblage. Other materials include dark red ccs (11 percent) and speckled gray ccs (10 percent), while translucent tan/gray ccs (1 percent), brown ccs (1 percent), tan ccs (1 percent), and green ccs (1 percent) occur in low frequencies. There are also four splintered stones of clear quartz and two of clear quartzite that may be manuports. All of the ccs materials are available locally in alluvium of Gyrshelun Creek and the Khilok River. The debitage assemblage (338 pieces) includes 7 cores, 66 cortical flakes, 179 flakes, 30 blades and blade fragments, 28 retouch chips, and 28 splinters. Cortical flakes (making up 18 percent of the lithic assemblage) include 37 primary flakes, 24 secondary flakes, and 5 fragments. These are typically made on dark gray and speckled gray ccs. Cortical flakes occur on every type of raw material present in the assemblage, further supporting the notion that raw materials were obtained locally. Further, the relative frequencies of raw material types are virtually the same for cortical flakes and noncortical flakes, indicating that unworked cobbles were carried to the site for reduction.

Core preparation and flake removal techniques were relatively expedient. The seven cores are informally prepared and include two monofrontal unidirectional flake cores made on cobbles of speckled gray ccs and dark gray ccs, a bifrontal bidirectional flake core made on a dark gray ccs cobble, a small end core (blades were struck from the end of the core rather than the face) made on a dark red ccs flake, a bipolar core made on a dark gray ccs flake, and two possible core tablets (platform rejuvenation spans) on dark gray and speckled gray ccs. Platform surfaces were simply prepared, with 84 percent of all cores and their removals having smooth platforms, and 11 percent having cortical platforms. Trimming and grinding of platform edges is evident on 55 percent of debitage pieces.

Blades and blade fragments make up 9 percent of the debitage assemblage. No large blade cores, however, were encountered during our excavations in 1996, but earlier excavations by Meshcherin in 1996 did yield one obvious blade core from component I (Goebel 1993). This is a unidirectional flat-faced blade core on dark gray ccs. Among the 21 tools, 11 are made on flakes or cortical flakes, indicating expediency in the production and selection of tool blanks. Nine tools are made on blades, and one, a chopper, on a cobble.

The presence of retouch chips in the debitage assemblage indicates that some secondary reduction activities also occurred at the Masterov Kliuch site. However, as with core preparation and blank manufacture, tool resharpening appears to have been expedient. Retouch invasiveness is minimal, with 14 of 22 tool edges having flake scars that extend less than 3 mm from the tool margin. Only two artifacts have retouch scars that are greater than 10 mm; these include a cobble chopper and denticulate. The 22 tools in the assemblage include 8 retouched blades and blade fragments, 5 retouched flakes, 2 knives, 2 denticulates, 1 of each of the following: graver, notch, cobble chopper, possible burin spall, and combination tool. Six of eight retouched blades are bilaterally retouched. Both knives are cortically backed, one on dark gray ccs and the other on dark red ccs. The graver has retouch that alternates between dorsal and ventral faces. The combination tool is an end scraper-knife on a dark red ccs cortical flake. Faunal remains from component I (1991 and 1996 excavations) number 18 pieces. Identified taxa include horse/ass (Equus sp.), marmot (Marmota sp.), and large mammal.

Component II

No features were encountered in component II, and, as described above, this component is considered to be redeposited and in a secondary position. The artifact assemblage from this component includes 104 lithic artifacts, 1 ceramic sherd, and 2 small unidentifiable bone fragments. The single ceramic sherd is an undecorated dark gray body shard similar to those described for component III, and is probably intrusive from that overlying stratum. The lithic artifact assemblage is made up of 81 debitage pieces and 23 tools. Raw materials include dark gray ccs (45 percent), speckled gray ccs (29 percent), dark red ccs (16 percent), translucent tan/gray ccs (8 percent), tan ccs (1 percent), and fine-grained gray ccs (1 percent). All but the last two raw materials can be found in local creek and river alluvium.

The debitage part of the assemblage includes 2 cores, 23 cortical flakes, 40 flakes, 8 blades, 1 retouch chip, and 7 splinters. Both of the cores are small bipolar cores manufactured on translucent tan/gray ccs. Among the cortical flakes are 13 primary flakes, 9 secondary flakes, and 1 cortical flake fragment. Even though cores are for the most part absent from the assemblage, the high incidence of cortical flakes (28 percent of the debitage) indicates that primary reduction activities occurred frequently at the Masterov Kliuch site. Among 42 platforms scored, 17 percent are cortical, 76 percent are smooth, and 7 percent are dihedral, further indicating that minimal platform preparation was involved in the manufacture of these artifacts. Among blades, there are three proximal blade fragments, four medial blade fragments, and one complete blade. All of these are made either on dark gray or speckled gray ccs.

Among the 23 retouched artifacts, there are 7 retouched blades, 3 retouched flakes, 3 side scrapers, 2 notches, 2 denticulates, 2 possible burins, 1 cortically backed knife, 1 graver, 1 end scraper, and 1 pointed tool. The retouched blades include one unilaterally and six bilaterally retouched pieces. The three side scrapers include a dejete scraper made on a dark gray ccs blade, a unilaterally retouched side scraper made on a speckled gray ccs cortical flake, and a side scraper fragment on a speckled gray ccs flake. Among the burins is a dark gray ccs blade fragment with a possible laterally burinated edge, as well as a dark gray ccs flake with a possible transversely burinated edge. The single pointed tool is made on a translucent tan/gray ccs blade fragment, and is dorsally retouched along both lateral margins. Retouch invasiveness is relatively low, with 17 (68 percent) of 25 measured tool edges displaying retouch scars that travel less than 4 mm from the tool margin.

Component III

While not a focus of our study, excavations in 1996 uncovered an intact Bronze Age living floor with two preserved features: an unlined hearth and stone-lined pit. The hearth, occurring in square 25S at an elevation of 76-82 cm below datum, consists of a wood charcoal and ash stain in an elongate oval shape, roughly 40-60 cm in diametre and 5 cm thick. A sample of the wood charcoal from this hearth yielded an AMS ^{14}C age estimate of 2895 [+ or -] 45 B.P. (AA-23648). The pit feature is situated in squares 25P and 26P, about 2.5 m east of the hearth feature. This 30-cm-deep pit is shaped like an inverted cone, with the top of the pit measuring about 100 cm in diameter, and the base only 10 cm in diameter. The pit's fill is an organic-rich loam with occasional charcoal flecks, small bone fragments, ceramic sherds, microblades, and several large stones.

The component III assemblage consists of 564 lithic artifacts (10 of which are tools), 23 ceramic sherds, and 12 small unidentifiable bone fragments. Retouched artifacts include four retouched microblades, two retouched blades, two end scrapers, one notch, and one hammerstone. The ceramic sherds are all of the same type, but appear to represent at least two different vessels. These are poorly fired, dark-gray colored ceramics that range from about 4 to 6 mm thick. Decorations include bands of diagonal incisions that consistently measure about 10 mm long. Similar pottery styles have been identified at other sites in the Transbaikal with late Holocene components, including Studenoe, Ust'-Menza, and Altan. Radiocarbon ages on such sites range from about 3500 to 2000 B.P., and are commonly attributed to the Bronze Age.

The Masterov Kliuch site contains two stratigraphically distinct early Upper Palaeolithic components. Component I has been accelerator radiocarbon dated to about 32,500-30,000 B.P., while component II has not been radiocarbon dated and appears to lie in a secondary context. Nonetheless, it too can be tentatively attributed to the early Upper Palaeolithic given technological and typological aspects of its lithic assemblage. Lithic assemblages are characterised by blade and flake primary reduction technologies, with blade cores being either flat-faced or "end" cores (but not prismatic). Bipolar reduction strategies are also evident. Secondary reduction technologies include unifacial as well as burin techniques. Tool assemblages include retouched blades and flakes, end scrapers, gravers, burins, knives, denticulates, and notches. Component II also yielded a small unifacially worked point on a blade.

The core technologies and tool forms present at Masterov Kliuch are characteristic of the Siberian early Upper Palaeolithic, dated elsewhere to between about 42,000 and 30,000 B.P. at sites like Kara-

Bom, Makarovo-4, Malaia Syia, Varvarina Gora, Kamenka, and Tolbaga. These sites in turn represent a widespread complex of flat-faced core and blade industries that spanned inner Asia from Uzbekistan in the west to the Transbaikal and perhaps inner Mongolia in the east during the mid-Upper Pleistocene, perhaps signaling the spread of anatomically modern humans from southwestern Asia.

Masterov Kliuch and Site Formation Processes in Siberia

Nearly all early Upper Palaeolithic sites known from Siberia (e.g., Kara-Bom, Kamenka, Sannyi Mys, Varvarina Gora, Tolbaga) occur in cryoturbated colluvial deposits; thus the geoarchaeological lessons learned at Masterov Kliuch have implications for these sites as well. Through careful excavation, three-point proveniencing of artifacts, conjoining artifacts, and measuring of trend and plunge of large artifacts found in situ, we were able to distinguish different degrees of integrity for the two early Upper Paleolithic components at Masterov Kliuch. Component I is characterised by tight vertical concentration and clustered horizontal distribution of artifacts suggesting a primary context, while component II is characterised by dispersed vertical and horizontal distributions of artifacts suggesting a secondary context.

Further, while we were able to refit some artifacts from component I, no conjoinable artifacts were found in component II. Trend and plunge of artifacts in these components are quite variable, but few artifacts were vertically oriented, suggesting that frost-heaving had not significantly displaced artifacts. Similar geoarchaeological studies are needed at sites like Kara-Bom and Makarovo-4 where artifact concentrations are thought to represent intact early Upper Palaeolithic living floors, and at sites like Tolbaga and Sannyi Mys where rings of stones are interpreted as dwelling features. Our experiences at Masterov Kliuch tell us that the behavioural context of artifacts at these sites, also situated in colluvial sediments along relatively steep slopes, could be disturbed, and that the putative dwelling features could be the product of natural, not cultural, processes. Clearly, reconstructions of early Upper Palaeolithic site structure and settlement behaviour need to proceed with careful consideration of geologic site formation processes.

Masterov Kliuch and Raw Material Procurement in the Siberian Early Upper Palaeolithic

The analysis of the Masterov Kliuch lithic assemblages, although based on a relatively small sample, provides an interesting glimpse into early Upper Palaeolithic raw material selection and procurement. In component I, all lithic artifacts recovered in our excavations are made on raw materials that are available in nearby alluvium of Gyrshelun Creek and the Khilok River, within 2-3 km of the site. Debitage analysis suggests that early Upper Palaeolithic flintknappers carefully selected fine-grained cryptocrystalline-silicate nodules from these sources and carried them to the Masterov Kliuch site for flaking. Core preparation and blade and flake manufacture occurred on the site, as did tool use, resharpening, and discard. There is no evidence of finished tools being transported to the site from some other location, or of exotic raw materials being brought to Masterov Kliuch from more distant sources.

Together, the evidence from component I suggests that early Upper Palaeolithic people were provisioning the Masterov Kliuch site exclusively with local raw materials for the manufacture of stone tools. The site, however, does not appear to have served solely as a task-specific quarry. Instead, evidence for multiple technological activities beyond those expected to be found at a quarry suggests that Masterov Kliuch served more as a residential base than a specialised resource extraction site. Further, the provisioning of this place with only local resources (gathered within 3 km of the site) and the apparent absence of exotic resources indicate that raw material procurement was "embedded" within other foraging activities that were carried out in the immediate area surrounding the site. This pattern of raw material procurement has been noted at other early Upper Palaeolithic sites in Siberia. Most early Upper Palaeolithic sites, including Makarovo-4 in the upper Lena Valley, Arembovskii in the Angara Valley, and Kara-Bom in the Altai Mountains, are situated very near sources of abundant fine-grained cryptocrystalline silicates. In all of these cases, greater than 95 percent of all finished tools are made on local raw materials procured within 5 km of the sites, and the full technological sequence of primary and secondary reduction is represented for these local materials (Goebel 1993, 1999). Further, exotic raw materials are absent from these sites. Like at Masterov Kliuch, then, the lithic assemblages from these early Upper Palaeolithic sites represents embedded raw material procurement strategies that focused on "hyper-local" lithic resources.

Masterov Kliuch and most other Siberian early Upper Palaeolithic sites further appear to represent hunter-gatherer camps that were repeatedly occupied, perhaps because of their proximities to high-quality raw materials as well as because of their ecological settings in areas of high topographic relief and environmental zonation, where diverse animal and plant resources would have been regularly available (Goebel 1999). This pattern of early Upper Palaeolithic raw material procurement and settlement suggests that early modern human hunters of northern Asia were often "tethered" to locations on the landscape where lithic raw materials as well as diverse faunal resources were locally abundant and accessible.

Similar patterns of raw material procurement and settlement have been documented for the Mousterian of southwest Europe (Mellars 1996) and the earliest Upper Palaeolithic complexes of Central Europe (i.e., the Bohunician and Szeletian). In these areas, the transport of exotic raw materials and the more logistical procurement strategies that they represent did not appear until the emergence of the Aurignacian after 35,000 B.P. In Siberia, such behaviours appear to have emerged even later, sometime after 25,000 B.P. during the time of the "Mal'ta Culture," the region's middle Upper Palaeolithic complex (Goebel 1999). Perhaps this means that aspects of logistical organisation and planning, so commonly portrayed to represent modern human behaviour, were relatively late in developing among the Upper Palaeolithic hunter-gatherers of northern Asia.

3

Fishing Pond and Farm Pond Construction with Biologists

We are Biologists who design natural farm ponds fishing ponds and clear water swimming ponds by using creative scientific techniques and practical experience in creating the most cost effective farm ponds possible from a recreational, economic and aesthetic spect. Forethought and Knowledge are our primary tools. This website is full of practical and technical information to help clarify your understanding of farm ponds and their challenges of:

- Optimising farm pond construction cost
- Farm pond water quality and nutrient issues
- Farm pond salutation
- Preventing flood damage
- farm pond dam safety and cost reduction
- Weed control in farm ponds
- Farm pond water supply issues
- Farm ponds for fishing ponds and other recreational uses
- Fire prevention ponds
- Maximizing the life of a farm pond.

Thoughtful farm pond design is going to address each of these issues to provide the most cost effective farm pond for a clean water supply, quality recreation and the longest possible life of the pond. A quality farm pond design will prevent pond dredging, dam rebuilding and most ongoing farm pond maintenance issues. There are way too many poorly designed farm ponds that require dredging every decade.

We design farm ponds to last lifetimes without dredging. When we discuss farm ponds, we are usually talking about a less demanding set of goals than our other work in swimming ponds, trout ponds, landscape ponds and fishing lakes where we sharply focused on investing in water clarity and aesthetics. In designing farm ponds we are not as demanding of our water supply because high quality well water or spring water is rarely available for most farm ponds. We do design farm ponds for clean safe water; they are just not as crystal clear as our other pond and lake designs.

Building a fish-friendly stream crossing requires an understanding of not only fishery biology but also basic engineering and construction principles as well as fluvial geomorphology (that is the processes that form and maintain stream habitat). Our intent in developing this web site was to provide the tools needed to plan and construct a fish passage structure that will be efficient, safe and conducive to fish passage. The procedures we suggest were derived from best practices recommended by public and private agencies and were selected to provide practical guidelines for designing long lasting, stable road crossings that will have minimum adverse affect on fish and their stream habitats. A number of state, federal and municipal agencies may have jurisdiction and permit authority over construction activities on streams and wetlands. Any person involved with planning, designing and constructing a stream crossing improvement should consult with the appropriate federal, state and local agencies from a very early stage within the planning process and comply with permit and construction regulations.

*E*arly in our history, rivers ran wild, and fish followed them according to their needs. This was a time when our fishery resources seemed abundant and without end. Since then, millions of culverts, dikes, stream diversions, dams, and other artificial barriers were constructed for transportation, to impound and redirect water for irrigation, flood control, electricity and drinking water. Many of these alterations have changed the open access features of rivers and streams and have taken their toll on our fishery resources. Improperly designed or damaged stream crossings are by far the most common cause of these problems.

Restoring Fish Passage Benefits people, Fish and the Environment

*T*he vast majority of crossings are culverts or bridges on small streams. Nearly all culvert installations are intended to serve the purpose of providing vehicle access, however, many also have an unintended function- they can block the migration of fish up or down streams. They also alter the geomorphic processes by which river channels form and maintain habit over time. For example, they may block or constrict the passage of sediment and large wood being transported by the river, and prevent channel migration.

Fish movements within streams are vital for maintaining healthy populations. Spawning migrations like those made by trout and salmon may be the most visible and dramatic, but seasonal (or even daily) movements upstream or downstream to find food supplies, refuge from predators, preferred temperatures or cover may also be critical.

Where fish habitats are divided into small segments (fragmented) by man-made barriers, whole populations may be eliminated, reduced or genetically damaged through the effects of isolation and inbreeding. The consequences can be disastrous for fish populations and other aquatic organisms within a watershed. It is a fundamental fact, fish need to move. For example, many fish need to move between feeding and spawning areas and make other seasonal movements to important habitats.

Impact of Fishing on Fisheries Resources

Technology and Fisheries Legislation

Research and Development are expensive and those investing in such activities seek to capitalize on their work and look to the law to protect their interest. In this respect, the export or import of technology is often controlled by government decree and there are many examples associated with trans-national corporations, classified(military) technology and where its importation may give an unfair advantage to one or a limited number of local manufactures. Likewise, whilst patent laws, protect the interest of an inventor, they also provide a vehicle for control over its use as well as further development of the invention.

However, research and development is also concentrated to a great extent in developed countries. In fact, conservative estimates by UNIDO put the contribution of developing countries in this respect at no more than 6 percent (less if china is excluded). Consequently, many developing countries see a growing need for a new approach to international transfer of technology, particularly in the course of implementation of UNCED's Agenda 21. Recent trends have shown a greater interest in technology acquisition by developing countries and some have elaborated acts in this regard in response to their desire to promote and stimulate scientific development, research and technological capabilities, the precautionary aspects of which need consideration.

The following sections will address first the general issue of regulating fishing technology for management purposes and then present some thoughts about precautionary approach to such regulation, before offering, in conclusion, some guidelines about implementation.

The Regulation of Fishery Technology

In general, capture fisheries benefit greatly from developments in technology arising from non-fisheries based industries. This is the case, for example, with regard to research and development in naval architecture, marine engineering, electronics and textiles without which there would have been little development in fisheries. Thus, laws promulgated for the purpose of fisheries conservation and management intended to restrict the transfer or development of technology, would not influence basic industrial research and development in the above mentioned disciplines. On the other hand, the fisheries sector does influence research and development to be directed towards aspects of capture technology by virtue of the market potential of the industry.

Finally, the fisheries sector also causes applied research to be carried out, for example in the case of fishing gear design and properties (e.g., fuel efficiency, selectivity). A prime example of this being the development of Turtle Exclusion Devices (TEDs) in shrimp trawls; once the technology had been proven. legislation followed requiring its use.

Technology Regulation: Subject to the provisions of UNCLOS 1982, each State may set conditions for the exploitation of stocks occurring in waters over which it has jurisdiction. Such conditions may also be applied to fishing methods and fishing materials. There is, therefore, ample license for a State to regulate the level of technology to be associated with the harvesting process.

In practice, many attempts have been made through fisheries legislation to restrict the importation of technology or to set limits within which a technology may be used, although these attempts have not always given the desired results. In fact efforts to restrict vessel sizes or power led to the development of "rule beaters" by the industry and some manufacturers simply rewrote their specifications to suit the law[1]. There are also many examples of technology having been held at a low level due to the general state of the development of the country concerned or for reasons of national security, trade agreements or labour considerations.

However, given that the massive rate of increase in landings from the 1950' was attributed to the development of new technology and geographical expansion of fisheries facilitated by such development, few countries would appear to have placed too many restrictions on its adoption. This fact, combined with the growth rate of the world's fleets, led to the present situation where the overall fishing capacity is clearly out of proportion to the available living resources of the seas and inland waters.

Given also that future trends could reflect a decrease in landings from capture fisheries if management fails to improve, fisheries managers should take into consideration technical developments in fleet restructuring exercises and in doing so, they should evaluate the *risks* associated with the adoption or non-adoption of new technology, as the case may be.

Technology, Safety and Risk: Fisheries managers and legislators must first of all consider the risks associated with using the law to control the adoption and use of technology as a tool for the application of the precautionary approach to management of fisheries. Such risks arise,*inter alia*, from the fact that technology is developed and adopted not only to improve fishing efficiency but also to improve safety in maritime traffic, to comply with international conventions and laws on labour, and to enhance the well being of fishing communities. Regulations aiming at controlling or limiting the use of technology may also be in contravention with wider laws related to technology transfer, where, for example, a government has introduced basic acts to enhance the technological capability of the country.

The inherent risks associated with maritime traffic have led to the development and availability of technology on the basis of which it has been possible to elaborate international conventions. Those countries ratifying such conventions legislate accordingly, requiring vessels to carry certain types of equipment and to carry out procedures that depend on associated hardware (and software). In such cases, the associated technical specifications for equipment are also internationally agreed. Furthermore, the time frame for the adoption of the technology under a convention, may take into consideration the age of ships, their size, area of operation, as well as the special needs of developing countries. For this reason, at any given moment in time, there is not a levelling out of technology in use world wide.

Within such international conventions, the safety of life at sea and protection of the marine environment play an important part and they set standards and regulations that can often only be met through the adoption of new technology and the bigger the fishing vessel, the more this is the case.

In addition, with respect to fishing vessels, such technology usually can be put to good use as an aid to fishing operations as in the case of the echosounder and navigation equipment, adding fishing efficiency to vessel safety.

In addition, many technological developments have been brought about through the need for increased efficiency, less crew, easier and safer working conditions and as a means to prevent marine pollution. Consequently, attempts to fishing vessels of some of their technical aids with the view to reduce fishing efficiency and capacity, may effectively render them uneconomic to operate and it could be illegal to do so with respect to national laws of those States that have ratified various international or regional conventions.

Sociological Analysis in Fisheries

Experience in fisheries, as in other fields of human activity, has illustrated the crucial importance of understanding the sociological factors at play in order to manage change in the use of fisheries resources. There are several ways in which the sociologist and anthropologist can assist the work of managing fisheries.

Understanding and Analysis of Context, Mechanisms and Institutions

Any planning or management activity should clearly be based on a detailed understanding of existing circumstances in the area or field which is being planned for. The sociologist has a key roll in providing this basis for planning and management by ensuring that the sociological context of the fisheries system in question is fully understood. This is a crucial complement to the understanding of the biological, economic and technical aspects of the resource as these sociological aspects can play as important a role in determining the ways in which people use the resource. Sociologists can assist in establishing the priorities and needs of people and institutions concerned with the resource. Subsequent plans can then attempt to identify those uses of the fisheries resources available which correspond to those priorities and needs. Likewise, the appropriateness of technological solutions or investment options for the exploitation of resources can be assessed with reference to people's ability make use of technology.

The role of the sociologist in the analysis of existing conditions within the fisheries system can include:

1. identification of the motivations and priorities of resource users
2. social institutions and their influence on resource use
3. analysis of institutions concerned with resource
4. analysis of leadership and decision-making
5. flows of resources within the community
6. roles of women, children and old people in the fisheries system
7. Place of different fisheries activities in communities and household livelihood strategies
8. modes of participation among difference groups of resource users
9. understand existing knowledge of resources among target groups
10. analyse of distribution of poverty and vulnerability among resource users
11. needs analysis in stakeholder communities

Designing Appropriate Interventions

The sociologist also has contributions to make during the more detailed design of those inventions and the formulation of plans, whether for fisheries development or management.

Among the key functions which sociologists can perform at this stage are:

1. identifying appropriate inputs - technologies, management measures
2. designing appropriate mechanisms for implementation of interventions
3. identifying or designing appropriate institutions for involvement in and management of interventions
4. forms of intervention which allow for and facilitate full participation by concerned resource users
5. identifying target groups and beneficiaries
6. designing mechanisms to incorporate gender concerns

7. designing mechanisms to incorporate age concerns
8 designing mechanisms for monitoring and evaluating social and economic impacts on different social groups.

Predicting Social Impacts

Once intended or possible fisheries interventions have been more clearly defined, the sociologist's main concern will be the identification of potential impacts deriving from planned interventions and how these are likely to be distributed among and affect different social groups.

Here the sociologist's contributions can include:

1. identifying and analysing impacts on particular target groups
2. identifying impacts on vulnerable groups within the community such as women, children, old people, refugees or the very poor
3. identifying and analysing the impacts of interventions on poverty
4. risks generated by social context - conflicts, lack of co-operation or participation by stakeholders
5. identifying analysing impacts on institutions and decision-making mechanisms.

Monitoring and Evaluation

1. Identifying appropriate indicators for social impacts
2. Identifying and designing mechanisms for monitoring and evaluating social impacts
3. Developing mechanisms for participatory, monitoring and evaluation.

Compensation and Mitigation

With the increasing importance of fisheries management as a focus of fisheries interventions, as opposed to the development of production, there is often a need within fisheries interventions for appropriate means of compensating stakeholders in the resource for the loss of access to fisheries. Here sociologists have a major contribution to make by determining what constitutes "appropriate" in the specific social and cultural contexts encountered. This can include:

1. design of measures to mitigate negative impacts
2. design of compensation packages
3. design of forums and institutions to support impacted groups

Risks Associated with Omission of Sociological Analysis

At the most basic level, the inclusion of proper sociological analysis as part of the process of managing fisheries will ensure that a whole range of social impacts are taken into account which might otherwise have been missed. These social impacts can often jeopardise the success or sustainability of fisheries interventions. Technologies which seem to fit the requirements of the resource and to offer new possibilities to fishers may be inappropriate because they require forms of organisation of the workforce which are incompatible with social norms or work traditions. Efforts to promote management of resources can fail because they attempt to make use of institutional mechanisms which seem relevant to resource management but are actually more concerned with other aspects of the social and cultural life of the community.

Assumptions About Resource Users

The omission of a sociological analysis can result in many assumptions regarding the motivations, interest and priorities of resource users finding their way into management plans and project design in fisheries. It will often be assumed that the "subjects" of management or planned interventions will have the same understanding of those interventions as managers or planners themselves. But the differences in viewpoint between those actively engaged in exploiting the resource and making a living from it are often profoundly different from those concerned with planning for the development or sustainability of a resource. If the assumptions of the planner are accepted without reference to resource-users themselves, there is a high probability that they will reflect the priorities and cultural background of administrators themselves rather than those of target groups.

Stakeholder Participation

Any project or intervention planned by external agencies depends, for its success, on the participation of stakeholders. This can take place in many different ways ranging from simple consultations regarding interventions which have already been decided elsewhere to full stakeholder ownership of the entire process. Sociologists and anthropologists can play a key role in looking at the process of participation and determining the best ways in which participation can be ensured. Failure to do this can carry a variety of risks. Too much of an emphasis on stakeholder participation, or a purely rhetorical commitment to participation on the part of administrators and managers, can lead to false expectations and disillusion when they are not realised. However, where the environment is right and the opportunities and mechanisms for stakeholder participation exist, sociologists can push for greater participation by stakeholders and so ensure greater sustainability for the project as a whole.

In management programmes this can be absolutely essential. Where people are being asked to renounce current resource use for relatively uncertain future benefits, a prolonged process of analysis and consultation is often required beforehand in order to reach a consensus on management action. Often, an important role for the sociologist here will be in identifying the discreet social and cultural groups who need to be "represented" during such consultations. In management activities, it only takes one group of resource users to be "missed out" and so feel that their interests are not being taken into consideration for the entire process to falter.

Limits of Sociological Analysis

The inclusion of sociological analysis in analysis of fisheries systems as a whole is increasingly recognised as being of crucial importance. However, the limits of such analysis, particularly within the context of processes of management and planning as they are often encountered within institutional and political frameworks, also need to be recognised.

Complications as a Result of Sociological Analysis

Sociological analysis attempts to identify and understand the various factors generated by the social setting which affect the ways in which people behave. For those involved in managing fisheries, the concern is obviously to understand how the various mechanisms at work in a particular society - for example ways in which people behave which are considered "normal" (norms), social institutions such as marriage or the ways in which people aggregate to carry out certain activities - affect their interaction with fisheries resources and, conversely, how changes in the fisheries resource will affect those social mechanisms.

The inclusion of this sociological focus results in a notable complication of the process of determining how to manage change in fisheries. Sociological analysis has to be relatively comprehensive because it expands over a wide range of fields and variables which can all have a significant influence on people's interaction with a resource or group of resources. As indicated in the discussion of fisheries systems above, this analysis cannot limit itself only to those people who use fishing technology to exploit fisheries resources. A much broader range of human activities connected to the resource has to be considered in order to be comprehensive. Besides those who catch and produce fish, there are those who process it, market it and all those whose livelihood depends on services to the sector.

In addition, fish has value primarily as food for people and the nutritional aspects of fisheries are part of the human element in the system. The numbers of people who depend in some way on fish as food may go far beyond those immediately engaged in the fisheries to include a vast number of local and distant consumers. Any of these people could be affected by changes in the state of resource and the technology available for their exploitation.

Conflict Generation and Resolution

The processes in which fisheries managers and administrators intervene can often generate conflict. This can range from the conflicts caused by perceived advantages acquired by one group over another as a result of the use of new technology to conflicts over access to increasingly rare aquatic resources. These conflicts can be the direct result of uninformed project intervention or they can be the result of other factors which may only be catalysed by changes introduced by external interventions in the sector. Often, it falls to sociologists and anthropologists working with fisheries teams to "deal" with such conflict as they are perceived as having a better understanding of its causes. At least where proper analysis has been carried out beforehand, sociologists should indeed have a better grasp of the dynamics of conflicts which focus on fisheries resources and their use. But this understanding alone will not necessarily assist in the resolution of such conflicts.

Resolving Disputes In Community Resource Management In Africa

In a recent project concerned with National Park management in an East African country, plans to introduce community wildlife management in areas immediately surrounding the park, giving local communities significant control over the wildlife resources in their areas, generated considerable conflict. While most local people living in the adjacent area supported the idea, a wide constituency of other stakeholders were opposed to it. Local conservationists perceived it as a threat to the park as local people had, in the past, been involved in poaching within the park.

Hunters from nearby towns regarded access to wildlife in the areas outside the park as a right and viewed the concept of control by local villagers with great suspicion. The resolution of this impasse was eventually entrusted to an anthropologist with long-standing local experience and knowledge of the area.

His experience was crucial in being able to lobby all the concerned stakeholders and eventually ensure the establishment of mechanisms to resolve disputes and overcome the widespread suspicions created by the introduction of a new form of wildlife management for that area. Sociologists may be able to help by identifying key political or traditional opinion-leaders who may need to be involved in resolving such conflicts. However it would be misleading to assume that their skills necessarily include the placating of irate stakeholders as a result of misguided planning interventions.

In most circumstances, sociologists' and anthropologists' best contributions can be made in preventing or at least predicting such conflicts. Resolving them generally has to be an activity led by the concerned stakeholders themselves and their leaders and institutions. This does not preclude a role for outsiders, including sociologists or anthropologists involved in fisheries teams, as intermediaries in such conflicts. The experience related in Box 1, which is not specifically related to fisheries, describes how the experience and local knowledge of an anthropologist or sociologist in particular circumstances can be of key importance in resolving such disputes.

Managing Change for People

People, both as individuals and groups, are extremely complex subjects for any kind of analysis. In economic analysis, at least some of the variables taken into consideration are quantifiable and relatively predictable, although just how predictable is subject to considerable dispute. By contrast, many of the subjects of sociological analysis are very difficult to even identify and impossible to quantify or predict in any way. Factors such as the motivation of individual fishermen to take risks or people's perception of the marine environment are the result of different combinations of highly dynamic elements.

The culture in which people live, its historical development and the outside influences to which it has been subject may play an important role in determining how people view risk and what elements enter into taking decisions about fisheries. Current and past economic conditions can clearly play an important role as well. The social and political structure of the community at any particular moment may also influence which risks are taken and who takes them.

At another level, each individual's family and its background, the age of their children, what they know, believe and what they think is important, their access to information (which includes both their formal and informal schooling and what they saw on the television last night), the prices they receive for different types of fish and what they have to pay for the various items they consume every day such as food, fuel, clothes, cigarettes are all likely to contribute to the decisions which they take about resource use.

The complexity of the various interactions which constitute the sociological element in any system makes proper analysis a daunting task and one which can rarely be carried out satisfactorily in the context of a "normal" planning process.

The kind of research required to truly be able to understand conditions from the point of view of other people living in a different culture requires more time than is generally available during Managing change procedures. This means that sociological analysis during planning for projects or programmes usually has to rely heavily on secondary sources and on relatively limited primary collection of data.

Sociological Analysis for Fisheries - A Framework

Structuring Sociological Analysis within the Process of Managing Change in Fisheries

Some of the contributions which sociologists can make to the process of managing change in the fisheries sector were briefly described and the importance of including the analysis of these areas in the process discussed.

In some of the key elements and concepts in sociological analysis are presented so that fisheries administrators and managers, and other non-specialist colleagues working with social scientists, will understand better what it is that they are concerned with and how they go about analysing social issues. Different specialists in sociology and anthropology will inevitably have different approaches to the analysis of particular situations and there are many different ways in which these key elements could be presented. The following framework suggests one approach to presenting these key elements in a structured way which may also assist fisheries managers in fitting the analysis into their work.

Levels of Analysis of Social Issues

In practical terms, sociological analysis in fisheries can be focused on a few key levels. These levels can be used to provide a basic framework to sociological analysis for fisheries.

Five "levels" of analysis are presented here:

1. Gender
2. Age
3. Community
4. Household
5. Production-unit

These five elements can generally be considered the "building blocks" of most social systems the world over so an understanding of how they are constituted and their significance is fundamental. Clearly, in every single culture there are important differences in the relative importance or emphasis given to each of these elements and, in some cultures, there may well be levels which need to be added to these.

For instance, in many areas of sub-Saharan Africa the tribe or clan can constitute a key social level which may need to be included in addition to the community and the household. But the five presented here can be considered of reasonably universal importance and a starting point for almost all sociological analysis. These levels are not discreet. There are many overlaps between them and understanding the interactions between levels can be critical in sociological analysis.

Gender

Gender issues are concerned with the complex of social, economic and cultural factors which distinguish men from women. Concern for the widespread gender bias in much development work has grown out of the realisation that, while development agencies the world over were mostly run by men and mostly dealt with men, much of the productive work in many societies was carried out by women and interventions which were considered good by men were not necessarily regarded as good by women. The focus on gender is concerned with both recognising and understanding the differences in roles between the sexes and with taking account of the very considerable differences which exist in gender roles between different societies and cultures. Plans for development in any sector must take account of those differences and not regard the impacts of development, whether in fisheries or other sectors, as "gender-neutral".

In many parts of the world, the social status accorded to women, and their relative lack of influence or participation in decision-making, makes them particularly vulnerable. The fact that most capture fisheries are carried out by men, particular in the commercial sector, means that the concerns of women in fisheries are easily ignored. But any changes in household or community circumstances will inevitably affect women as much as men and therefore full account has to be taken of women's priorities even where their direct involvement in the activities in question may be limited.

Apart from this, many key sub-sectors in fisheries are dominated by women. Much small-scale fishing and collection of other foods found in water may be carried out by women and constitute an important source of household food supply. Marketing and processing in both the artisanal and the industrial sectors in many parts of the world are predominantly carried out by women. In West Africa, the central role played by women in the handling of fish catches extends to the financing of artisanal fishing companies.

The consideration of gender issues is therefore a key element in the analysis of any fisheries system. Any omission of a gender focus in fisheries decision-making can be said to automatically mean that 50% of the required analysis has been simply forgotten. All interventions which have an impact on people's livelihoods or patterns of use of fisheries resources are likely to have significantly different impacts on men and on women respectively. Therefore the preparation of such interventions must take full account of the different outlooks and needs of both gender groups. Ensuring that managers and administrators have a full understanding of the roles, priorities and needs which are determined by gender is therefore one of the sociologist's basic tasks.

Gender Specialists: In many fisheries management activities there will be a justification for a specialist in gender issues who will concentrate entirely on this aspect. A gender specialist is likely to be a sociologist with experience in the analysis and investigation of gender issues and, in particular, how the impacts of changes in technology, resource use and production methods affect men and women differently. In almost any circumstance where relatively wide-ranging impacts are foreseen from a fisheries intervention, gender specific issues are liable to arise and these will generally require special skills in order to be understood. For example, almost any fisheries management initiative is liable to affect the flows of benefits within and between fishing communities and the impacts of any changes in benefits will be felt by fishing households and their members. In order to properly assess the impacts at the household level, attention has to be paid to the way resources and benefits are distributed within the household and this will generally require skills in intra-household research which not all sociologists will be able to fulfil. Gender-related issues are often complex and difficult to research.

In many societies there are major cultural obstacles which make it difficult for outsiders such as planners or researchers, and particularly male outsiders, to have sufficient contact with women in order to properly assess their needs and priorities and how they might differ from those of men. In addition, in many countries fisheries departments and fisheries agencies will tend to be predominantly staffed by men and the inclusion of a specialist in gender issues can go some way towards redressing this imbalance and ensuring that the concerns of women are properly aired and discussed during the preparation of plans for fisheries development or management.

Key Concepts:

Gender Roles: This refers to specific duties, modes of behaviour or activities within the household, the community, the production unit or society at large which are determined by gender. For example, in many small-scale fishing societies, capture fisheries using fishing craft and larger types of fishing gear are often considered to be "men's work" while the processing of fish catch and, on occasions, its sale are regarded as "women's work".

In such cases, there is a clear, gender-based division of labour in the productive process. Such divisions can take many forms. Migration for work may be seen as a male activity while women stay at home. In some cultures, men may be expected to earn income from their work while women's work primarily provides food for the household.

More generally, there tends to be a distinct gender-based division of roles regarding the maintenance of the family and household. The investigation and understanding of these roles is more complex than looking at roles in production because people's perceptions of these roles are often culturally based and therefore quite difficult for them to analyse themselves.

In some traditional rural societies, particularly in Muslim countries, men may continue to say that women do not work, even when it is very obviously not the case, because there is a cultural ideal that women should not work and it may be very difficult for men to actually "see" that women do work. Alternatively, women's work may not be recognised as "work".

In some Southern African communities, during the dry season, men may be practically unemployed and spend most of there time socialising and drinking while women look after vegetable gardens, collect wild produce for food and collect firewood and water. But even in conditions of such stark contrast, men may say that women do not really work because their work does not earn income.

Productive and Reproductive Labour: The different gender roles outlined above are often closely interlinked with another important concept in gender analysis, the division between "productive" and "reproductive" labour. "Productive" labour, as the phrase suggests, is labour which produces food or goods for the generation of income or the sustenance of the household. Fishing activities will almost invariably constitute productive labour, as will the processing and sale of fish, farming activities, and any labouring tasks which are remunerated either in cash or kind.

"Reproductive" labour includes those activities which do not "produce" an output which is directly converted into cash or consumption but which nevertheless contribute to the maintenance of the household and family. Time spent in child-rearing is one of the clearest examples of this form of labour but other important activities might be the collection of fuelwood for household use or the fetching of water. The divisions between productive and reproductive labour are not always clear-cut. Fuelwood collected for the household can also be sold or exchanged for goods as can water.

What is of key importance is that women frequently have a double burden of productive and reproductive labour and sociologists have to ensure that the patterns of women's labour in both areas is well understood and taken into account.

Just as gender constitutes a fundamental factor determining social behaviour, so age is also important as it has important affects on the needs and priorities of people. In the case of age, the divisions which can be established between different age-groups are not clear or universal; different age groups are categorised in different ways and according to different criteria in societies the world over. But generally every society has a clear age structure with each age group associated with certain types of activity and holding certain responsibilities.

As with gender, age is also an issue which cannot be approached with cultural preconceptions about what the roles and needs of specific age-groups might be. Urban, educated preconceptions regarding the ages at which children should be at school or working, or the specific needs and rights of older people may be completely inappropriate in the face of radically different needs and priorities among poor rural fishing households. Frequently an important role for the sociologist on fisheries teams will be to ensure that the team as a whole is able to "see around" their own preconceptions of this kind.

A good example might be the role of children in fishing in certain less developed countries. Fisheries specialists from an urban setting or from outside of the country might assume that children's fishing activity is not important or can be relegated as "play" because the catch is not sold or at least does not turn up in local markets. But, where children are frequently observed fishing, even if it is with relatively simple gear, it would be the sociologist's responsibility to ensure that the extent and importance of children's fishing is properly investigated. In areas such as rural Bangladesh the contribution of children to overall fishing effort as well as their contributions to household food supply, networks of reciprocal exchange outside the formal market and small rural markets could turn out to be extremely significant. However, most adult fishers and most fisheries officers would generally regard these children's activities as of peripheral importance.

A better understanding of the role of age in determining levels of economic and social participation may be of great importance when it comes to targeting interventions. Frequently, those concerned with fisheries development or management assume that their target group is made up above all of economically-active "adult" males. This assumption needs to be actively questioned by sociologists who have to take responsibility for checking possible impacts or concerns for other segments of the population which may be less "visible" and more difficult to talk to, but which may be of equal relevance in terms of their dependence on fisheries.

Vulnerability: One of the features of both the very young and the very old, particularly in poor, rural communities, is their relative vulnerability in conditions of stress or change. This can be caused by a wide range factors - lack of access to resources, susceptibility to disease, or cultural norms which give priority to working people when it comes to the distribution of resources and food. Whatever the dynamics of vulnerability among the young and the old, they are almost always among those most at risk when changes of any kind are introduced and sociological analysis has to establish how consistent those risks are and what the effect of decisions about fisheries resource use are liable to be on these groups.

For example, if access to resources is to be restricted as part of a fisheries management plan, leading to a reduction in earnings or livelihood opportunities, it needs to be investigated how these reductions might affect these vulnerable age-groups. Changes in resource access which may be easily absorbed and adapted to by adult workers may have more serious impacts on young people and the aged.

Rates of Dependency: The numbers of people dependent on active producers will affect the decisions made by those producers. Particularly among poor rural households, any risk as a result of changes in patterns of resource use will generally be avoided wherever possible. Where the household includes numbers of temporarily dependent people - young children or old parents - the incentive for changing behaviour is further reduced. This can have an important affect on the willingness of fishers to renounce resource use in favour of resource management.

Community

Definitions of Community: The "community" is often the most convenient administrative level at which to implement activities in the field. However, the real meaning of the community in different cultural, social and political contexts is not always fully appreciated. It is often assumed that people living within certain administrative boundaries, or living in close proximity to one another automatically constitute a "community" which has certain common interests, goals and a communal sense of identity.

This is deceptive. People's real sense of community may not be defined by residence at all but may have more to do with common occupation, common socio-economic status, and kinship or ethnic ties. For example, in some parts of South Asia, traditional marine fisherfolk may regard themselves as part of a caste "community" of fishers which includes people living at considerable distances along the coast. On the other hand, they may feel practically no sense of common identity or interest with agricultural communities located a few hundred metres away even though they belong to the same administrative "village".

In other circumstances, horizontal links to a community of fishers of similar social and economic status may have less real significance in terms of common goals and priorities than other, vertical links of patronage and economic or political influence.

In small-scale and artisanal fisheries, links between fishers and the dealers who buy their catch or moneylenders who provide finance for fishing operations are frequently of great importance. Attempts by fisheries development agencies to replace these vertical linkages with linkages, such as co-operatives, based on perceived common economic and social interests have often foundered as they have oversimplified existing patronage networks as being inevitably exploitative.

Sometimes these ties are highly exploitative but their social and cultural significance also has to be taken into account. These networks often ensure links between the producer and the outside world as well as a form of social security and flexible support for vulnerable fishing households subject to the vagaries of the resource and the seasons which co-operatives may find it very difficult to imitate.

At the same time, there may be important ties of kinship and mutual obligation between primary producers and "exploitative" middlemen which are not easily dismantled and replaced by more impersonal formal systems. Administrators and managers therefore have to understand the often over-lapping "communities" in which fisheries resource users live and the relative priorities accorded to different sets of community allegiance.

In fisheries management this is of particular importance as this analysis will define who the key stakeholders are in a particular resource. The sociologist has to map out these interlocking, and at times conflicting, community and stakeholder groups and clarify how they are defined, what holds them together and their relations to different sets of resources and their use. The internal structure and organisation of different communities also has to be taken into account. Even apparently homogeneous "communities" will have important internal divisions along lines of kinship, occupation or social status which may affect participation, decision-making and other areas of basic importance for development or resource management activities.

Centres of Decision-making: A key element in the analysis of community structure is the identification of centres of decision-making. These may consist of individuals, formal and informal institutions or groups where decisions are made. Clearly, beyond the identification of the centre or "locus" of decision-making, the sphere of influence of each centre of decision-making needs to be defined. Leaders indicated as having key responsibilities within the formal or official structure may have relatively limited influence over a specific range of aspects of community life. Informal or traditional leaders often exert greater influence over a wider range of spheres.

Decision-making in most communities, whether in remote rural areas or more developed suburban communities, tends to develop into a complex pattern of overlapping responsibilities determined by economic position, political influence, traditional beliefs and custom, and kinship.Decisions regarding resource use are likely to be particularly important in fisheries. In many societies, the mechanisms for deciding on the use of common-pool resources, such as waterbodies or marine areas and the fish in them, are constantly changing.

The powers of the State to determine how such resources are in use may have not yet displaced traditional systems which empower community-level institutions with control of such resources. In some parts of southern Africa, formal control of resources over large areas may reside with tribal leaders who may be quite distant from the resources in question. Recognition of these different patterns of control is extremely important if conflicts and confusion over the relative responsibilities of local leaders, distant traditional authorities and formal government agencies are to be resolved.

Particularly in fishing communities, the centres of decision-making for land-based activities and those which take place on the water are sometimes quite different. Fishing activities often distance fishers from the decision-making mechanisms which govern life in their home villages and, at sea or on distant fishing grounds, quite different mechanisms for taking decisions may come into force. Along the coasts of West Africa, many fishermen migrate long distances on a regular basis, crossing international borders and residing for significant periods in foreign communities. Clearly, in such circumstances, fishers may move through the spheres of influence of various centres of decision-making regarding the resources which they exploit.

Household

Below the community level, the household is likely to be the next key unit in any social analysis, but considerable care has to be taken regarding what constitutes a "household" in different social and cultural contexts. In many urban areas or in some parts of the "developed" world, the household generally corresponds to the nuclear family, sometimes expanded to include some additional generations (grandparents or grandchildren) or some relatively close kin.

However, in many rural areas the household can consist of a wide range of combinations of kin or people connected by links of patronage and employment to the core household members or the head of household. Care has to be taken by those making decisions or assumptions regarding household-level impacts or interventions to be clear about what "household" means for each target group. It is the sociologist's responsibility to ensure that the various types of household encountered in a target area have been identified and the factors which determine household composition understood.

Marriage customs and patterns of inheritance will be of key importance in determining forms of household. Where polygamy is widespread, as in many parts of the Sahel and western Africa, households headed by individual wives of one husband may have considerable autonomy over certain aspects of production while husbands contribute to specific elements in the family's livelihood. Wives involved in fish processing in West Africa often have far more developed economic relations with other fishers than with their husbands. Patterns of inheritance are also significant in determining how households are constituted and how they evolve over time. Where norms of inheritance concentrate ownership of key fisheries technologies in the hands of one offspring at the expense of others, this will determine particular patterns of household composition once children reach a certain age. Children reduced to working for older offspring may have added incentives to move out and found their own households or migrate in search of other work. Patterns of inheritance can even become major determinants of mobility of labour out of fisheries into other sectors.

Inheritance of rights of access to fisheries or tenure over fishing grounds, or their transfer as part of the exchanges of goods and rights accompanying marriage, can influence overall patterns of fisheries exploitation in ways which fisheries managers must be aware of. In some areas of Melanesia, patterns of inter-island or inter-village marriage may lead to such complex patterns of reciprocal rights of access to fishing grounds that nominal tenure over marine areas actually has no real relevance in terms of hypothetical controls of fishing activity. So many different groups may have acquired rights to different fishing areas that, for all practical purposes, fisheries access is open.

4

Fish and Fishing Culture

Introduction of Integrated Fish Farming

Chinese integrated fish farming system is so broad in scale with so many models that there is no comparative in the world, and it has developed its own characteristics as well. This technology attracts the world's attention. Perhaps, the reason for this is that it can fully develop and utilise local natural resources and it can produce more materials for the people and it can satisfy the needs of the world economic development on many aspects.

Figure: *Integrated fish farming.*

The Advantages of Integrated Fish Farming

Establishment of a Man-made Ecosystem without any Wastes

In developed countries, the more their industry develops, the more intensive their farming is, and the larger the scale of chicken and pig raising factory farms is, the more the wastes accumulate, e.g. in Japan, only the excreta of cattle, pig and chicken reach more than 70 million metric tons per annum. If this great amount of excreta is not disposed of, it must pollute the environment.

The annual output of animal manures is considerably high — 10.4–11.8 tons per black-and-white cow; 4.1–4.7 tons per pig; 0.04–0.07 tons per duck. The decomposition of so much excreta produces gaseous ammonia, etc. The odour pervades everywhere. As a result, the water, the land and the air are all polluted and it will jeopardise the people's health.

In fact, the manures of livestock and poultry are good organic fertilizers for fish farming. About 40–50 kg of organic manures can be converted into one kg of fresh fish. If the farm combines fish farming with mulberry cultivation, sericulture and silk extraction from the cocoons, the pupae will be a fine feed for fish while the worm faeces and waste water obtained at the processing factory are used as pond fertilizers.

In an integrated fish farm, the manures of livestock and poultry and the dregs, lees and waste water obtained from starch processing and wine brewing are all used for fish farming. The pond silt can be used as fertilizers for terrestrial fodder crops which can in turn be used to raise livestock and poultry or directly for fish farming. Thus, it forms a recycling ecosystem, which utilises various wastes and afterwards, an agricultural ecosystem can be set up without any wastes.

Increasing the Food Supply for the Mankind

Short supply of food and protein is a serious problem that we are now facing. Therefore, merely using the pelleted feeds of grains and animal protein as fish feeds is not economical and often makes the food for the mankind more insufficient.

If the grains such as wheat are used in fish culture, the output could be up to 4815–9750 kg/ha. But the average Food Conversion Rate of grain is 3, i.e. 3 kg of grains in dry weight could converted into 1 kg of fresh fish. This kind of fish culture is not suitable for the countries which want for food. However, the natural food organisms cultured in fish ponds by using organic manures could totally take the place of pelleted feeds or grains. The quality of fish will not be changed by using animal manures as fertilizers in fish culture. The daily output could reach 15–32.25 kg/ha or even higher. E.g. Helei Fish Farm in Wuxi has not only built fish ponds with an area of 69.4 ha but also a dairy with 100 cows and pigsties with 1000 pigs and duck yards producing 10,000 eggs per day. It has become a subsidiary food production base, which is an integration of aquaculture industry and commerce. So far as aquaculture is concerned, they combine fish farming with livestock and poultry raising, however, putting fish farming first. In 1981, it supplied not only large amount of fertilizers to produce 600,000 kg of fish for market but also supplied 490,000 kg of pork, etc.

The scope of integration in an integrated fish farm could be considerably wide. Apart from fish cultured in the water body, the water surface can be used for goose and duck raising, pond dikes for fruit tree and mulberry cultivation or for setting up pig-sties, slopes for fodder crops. The products from an integrated fish farm are not only fish but also meat, milk, eggs, fruit, vegetables, etc. It is apparent that integrated fish farming can fully utilise the water body, the water surface, the land and the pond silt, etc. to increase the food supply for people.

More Job Offer

In unitary fish farm, the labour force cannot be reasonably employed while the integrated fish farm may offer more jobs, for example, in Helei Fish Farm, the occupations except fish culture offer 149 jobs, among which there are 48 persons for duck raising, 19 for cow farming, 14 for pig raising.

Reducing the Cost, Raising the Output and Increasing Economic Benefits

Nowadays, the problem that the aquaculture in developed countries is facing is that the cost of pelleted feeds is too high. It is related to energy crisis and protein shortage also. The integrated fish farm produces feeds and fertilizers for itself, savrer energy and reducing. Expenditure considerably, e.g. in 1981, Helei Fish Farm produced 5.5 million kg of pig manure, 1.85 million kg of cow dung, 1 million kg of duck manure, 9.5 million kg of waste water from silk extracting workshop.

The total amount of organic fertilizers was 17.65 million kg. Outflow of animal excreta by gravity could reduce the production cost by about 97 year each mu of fish pond. In 1976, Helei Fish Farm began to put duck, cow and pig raising into integrated fish farming system, food processing industry in 1979 and commerce in 1980. During the period of 1977–1981, the yield of fresh fish increased from 115,000 kg to 600,000 kg (the same area) while the yields of livestock and poultry increased from 135,000 kg to 490,000 kg. In 1981, average income per capita increased by 120. Over the year of the establishment of the fish farm in 1966. The annual net fish production of integrated fish farms, only by using organic manures.

The Characteristics of Chinese Integrated Fish Farming

The integrated fish farming practices have developed into a complicated structural network in line with the local conditions.

China is a vast country with a large population and varied natural environments. The agricultural structure and the economic conditions of each locality are also different. Therefore, various integrated fish farming systems in line with the local conditions have developed into a complicated structural network. The Pearl River Delta is located to the south of the Tropic of Cancer. Its annual solar irradiation is 110 kcal/cm^2 but averaging 60 Kcal/cm^2. Average temperature is about 22°C; 2–3 days of frost a year; more rain and high temperature in summer with maximum temperature of about 37°C and relative humidity 76–85%; annual sunshine time between 2000 and 2500 hours. Such geographic and climatic conditions are very conducive to the cultivation of mulberry trees, sericulture and fish farming. Hence, the farmers in the Pearl River Delta through production practices have been able to take advantage of local natural resources by integrating mulberry cultivation and sericulture with fish farming, which leads to the establishment of a complete, scientific "mulberry plot-fish pond" man-made ecosystem.

Even in the same geographical zone, the items of integration by various production units are different. Helei Fish Farm, Xinan Fish Farm, Helei First Fishery Brigade, Wangzhuang Fish Farm as well as the Municipal Fish Culture Farm in Wuxi practise different items of integration. Liutan village has recently incorporated agriculture, sideline occupations, aquaculture and commerce into its integrated system. The integrated fish farm was established in 1980. The farm site was earlier a water-logged paddy field of 17.33 ha, subject to annual flooding and with low yield of agricultural crops. Based on the topography of the low land, the farmers restructured the land into fish ponds of 10.67 ha. in the winter of 1979 and the spring of 1980. The farm consists of 15 grow-out ponds (8 ha), 11 fingerling ponds (2.67 ha). Apart from pig raising, green fodder planting is the major item of the integrated management. It has 1.33 ha of pond dykes for the cultivation of English rye grass (Lolium pereme), sow thistle (Lactuca indica), etc.

They utilised water cane shoots field to cultivate duckweed with average yield of 37,500–45,000 kg/ha. With sufficient supply of green fodder, Grass carp and Wuchang fish (Megalobrama spp.) become the dominant cultured species, with the yield of 5250 kg/ha, that is 45% of the total pond fish production. It indicates that the maximal efficiency of the integrated fish farming system can be obtained only through maximal utilisation of natural conditions and the agricultural characteristics of the regions concerned.

The socio-economic conditions should be taken into consideration in developing integrated fish farming. The fact is that the diversified economy develops in the harmonious interaction among the socioeconomic conditions, agricultural production, and natural conditions on the basis of regional climatical differences. Since the development of integrated fish farming is site-specific and the conditions of each site are different and complicated either from the macro or micro view points, the linkage between the trades of integrated fish farms in China become a complex matrix. Simple modele involving mono-integration of fish-cum-animal husbandry, fish-cum-poultry or fish-cum-crops are now getting fewer and fewer in China with the exception of small-scale individual farms.

Utilising Wastes by Various Ways of Recycling

The integrated system of fish farming is rather complicates involving various forms of integrations and methods in utilisation of organic wastes. The animal excreta alone can be effectively utilised through employing various methods and techniques.

1. Fresh animal manure can be applied directly to the fish ponds. Pigsties, poultry coops and pens for ducks and geese can be constructed on the dikes or above the ponds. Fresh manure thus enter the ponds directly, avoiding energy losses due to processing of manure and transportation. The feedstuff of livestock are not fully digested and then can be directly utilised by fish. Therefore, the number of animals should be compatible with unit water surface.
2. The residues and liquid part after anaerobic fermentation to produce biogas are used for fertilizing fish ponds.

   Compost after aerobic fermentation can also be used for fish culture.
3. Animal manure can be used indirectly through one or two trophic levels in a food chain such as growing fodder crop to feed herbivorous fish, producing earthworms or other animal-

feedstuff for carnivorous fish directly or used as part of the composition of pelleted feeds.

4. Poultry manure can be used to feed pigs and pig manure can in turn be used as fertilizers for fish pond.

Further Development and Research Needs

The prosperity of agricultural economy and the application of new technology stimulated and improved the production of integrated fish farming, e.g. the use of aerators and the mechanisation, to certain extent, raised production efficiency of the classical mode of Chinese fish culture. Although integrated fish farming has long been practised in China and rich experience has been gained through years of improvement and considerably high yields have been achieved, the scientific basis for some of the existing techniques are still awaiting clarification.

In order to raise the efficiency of integrated fish farming considerably, the biological basis of integrated fish farming must be studied, finding out the rule and developing aquaculture technology. In theory, two ecosystems must be clarified: First, the integrated fish-livestock-crop land-and-water ecosystem research should be carried out on its structure and functions so as to set up an optimal and harmoneous ecosystem on a perfect structure and with sound functions; second, the pond ecosystem.

The structural research concerns the biological interaction among fish, livestock and crops and their matched proportion. Material cycle and energy flow should be measured to elucidate the biological relationship and exact proportion of quantity among these production links. So far as the land-and-water structural ecosystem of integration is concerned, crops are producers; livestock and fish are consumers; aquatic organisms and soil organisms are decomposers. It is a rather complete ecosystem. In this system, benign circulation demands that there be no wastes in any link of this structure and the production be not hindered by lack of sufficient energy. For this reason, it needs a thorough research on fish pond ecosystem. Manureloaded pond is a semi-closed man-made ecosystem. Being put into fish ponds, animal manures begin to enter into the process of decomposition by bacteria. It's a complex process of pond dynamics in which animal manures will be converted into fish protein. It involves many factors such as physical or chemical parameters or biological factors in food chains. In order to investigate the rule of variation and intricate relationship between biology and non-biology, the research needs cooperation and common efforts of different scientists and specialists from multidisciplines such as aquaculture, ecology, botany, microbiology and chemistry.

Integrated fish farming is a low energy consumptive but high efficient aquaculture system. It's a right way to develop fresh water fish culture and also it's a strategic measure of developing Chinese aquaculture. Integrated fish farming will play a promising role in solving the problem of "hard to get fish".

Integrated Management of Fish and Crop Farming

Fish-cum-crop integration is the most ancient and popular pattern of integrated fish farming.

The Necessity and the Feasibility of Fish-cum-crop Integration

Why can we combine fish farming with crop cultivation? It's due to the demand of fish feeds and the excessive pond silt. On one hand, abundant silt deteriorates the pond water and yet on the other hand, it is one of high quality manures for agriculture to plant fodder crops which can in turn be used as feeds for fish. Therefore, pond silt is a link between fish and crops in this integration.

Formation and Function of Pond Silt

Large amount of feeds and manures are put into fish ponds year after year. There is a considerable amount of residues and manures which often sink to the bottom. Moreover, the excrements of fish and aquatic animals, the corpse of the aquatic, and alluvial soil incessantly descend to the bottom too. The sedimentary organic material decomposed by bacteria forms a great deal of humus, which in turn combines with sludge on the surface of the pond bottom to form silt.

The thickness of silt varies greatly with many factors. Even in the same pond, the thickness in different locations is different. The mean thickness formed in a high-yielding earthen pond with target yield of 500 kg/mu and with no slope protection is about 10–20 cm, which is about 50–95 m^3/mu/year or 100–190 tons in wet weight.

The proper amount of silt is beneficial to fertilize the pond water whereas the excessive amount is not so good to fish. Microbes propagate rapidly when large amount of feeds and manures are applied. The pH value will decline, the BOD will increase, and gases such as nitrites NH_3, H_2S, CH_4, PH_3, etc. and surplus nitrites will do harm to fish. The median tolerated limit (TLm) value of Silver carp and Bighead fry in 24 hours at the temperature of 25°C is 0.91 mg/l and 0.46 mg/l respectively.

Grass carp is more susceptible (Lei Xingzhi Et al 1983). Surplus nitrites will easily induce hemorrhagic septicemia of fish. Silt contains a lot of ichthyopathogen parasites and other harmful creatures. The more the thickness of silt, the more the deterioration of the pond water. It directly affects fish yields and thus, the excessive silt should be removed after the pond is drained.

On the other side of the coin, silt is a manure of high quality which contains several nutritive elements. In Pearl River Delta and Taihu Lake basin, the pond silt is often removed 3 to 6 times every year.

The nitrogen content of pond silt/mu/year is equal to 481 kg of ammonium sulphate. Furthermore, as pond silt is quite complete in nutrition, it contains not only nitrogen but also phosphorus and potassium. Generally, T! of silt is removed per mu per year and that is equal to 1.2 tons of N.P.K. fertilizers. The quick-acting component of silt is equal to 119 kg of fertilizers, which can also serve as additional manure besides the basal one. Silt could increase the thickness of cultivation layer, improve the soil particle structure and strengthen its ability to absorb ions of N.P.K. and the ability to keep the water. It is beneficial to the prevention of fish diseases. Since it is also a slow-acting fertilizer which is beneficial to late crops. In terms of mass's experience, the quick-acting component of 100 kg of silt in dry weight can increase 1 kg of rice.

If we only use silt as fertilizer, 100 kg of silt in dry weight can produce 10 kg of rye grass. Provided that 50 m^3 of silt is removed from each mu of fish ponds, all the silt can be used to cultivate 6 mu of rice, and then it could increase about 500 kg of rice; if it is used to cultivate rye grass, the output can reach over 6000 kg/mu. The yield can be increased by 2.5 times. From investigation data, the yield of per-unit paddy field by using 10–15 tons of silt composted with green grasses is near to the one by using 5 tons of animal manures composted. The cost of digging silt in China is cheaper than the one of purchasing animal manures from outside the fish farm.

The Demand of Feeds in Fish Farming

The demand of both commercial feeds and natural food organisms is great in fish farming alone. The feeds and foods sometimes are hard to get and the great amount of supply also affects the balance of the market and ecology. Besides, the cost of transportation and energy consumption is surprising.

Figure: *fish aquaculture - fish farmer.*

The fish pond with the target net yield of 250 kg including 100 kg of herbivorous fish, 100 kg of planktoneater, 50 kg of omnivorous fish needs 1500 kg of aquatic grass, 1000 kg of vegetables, 150 kg of grains apart from manures. In order to meet the demand of feeds and foods in fish farming and to reduce the cost, it's necessary to combine fish farming with crop cultivation.

The Feasibility of Fish-cum-crop Integration

Aquaculture can provide large amount of silt and fertile water for agriculture. There still exists a potentiality of land on fish farms. E.g. the mean pond dyke is three meters wide, the gradient of a slope is 1:1.5–3. The mean area of a fish pond is 10 mu.

The arable area of pond dyke and slope and the area of the water surface is in the ratio of one to five. The arable land might be greater before May with shallow water in the pond. With extra 0.3 mu of forage field attached to one mu of fish pond plus all the areas available, the average forage field for one mu of fish pond could amount to 0.5 mu or even higher which can provide fodder crops for herbivorous fish and proportionable filter-feeding fish in one-mu fish pond with net target yield of 400 kg.

What's more, it's possible to plant aquatic plants on scattered unused surface. In short, it is necessary and feasible to integrate fish farming with crop production so as to fully utilise pond silt, arable land and water surface. As a result of that, the needs of fish feeds could be satisfied wholly or partially.

Major Patterns of Fish-cum-crop Integration

According to crop variety and various cultivation systems, fishcum-crop integration can be divided into four patterns: fish farming integrated with terrestrial crop production; cultivating crops in ponds first, fish farming next; fish farming combined with aquatic plant cultivation and dyke-pond system.

Fish-cum-terrestrial Crops Integration

All or most part of crops planted in the fodder crop field and corner plots on pond dykes and slopes are used as green fodder for fish and as fertilizers for ponds. This is the most popular pattern in fish-cum-crop integration.

Crop Variety

Choose crop variety which are palatable to fish, rich in nutrition, strong in resistance to diseases, easy to manage and has features such as well-developed root, which protects the slope, etc. If it serves as straw manure, it should be easy to decompose. Besides, the average yield of some leguminous plants such as Trifolium repens, T.pratense, Medicago sativa, Astragaluo sinicus can reach 5000–7500 kg/mu. These grasses serve as both feeds and fertilizers.

The average yield of some gramineous plants such as Pennisetum purpureum, Phalarosarundinacea, Pennisetum alopecuroidescross, P. purpureum can reach over 10,000 kg/mu. The seeds and young crops of other grains such as barley, wheat, maize and rice are also palatable foods of fish. The tender and juicy vines and leaves of sweet potatoes Ipomoea batatas and squash Cucurbita are pulverised as fish feeds for Grass carp in Wuxi area. The tuber and squash, etc. are cooked to feed feed-eaters.

Integrated Fish Farm

Animal Raising on an Integrated Fish Farm

The Purpose of raising animals on an integrated fish farm is to develop integrated fish farming and fully utilise the limited feedstuffs. The multi-stage utilisation of feedstuffs and fertilizers makes it possible to supply the community with more produce and to increase the income for the fish farm as well. Such activity in China is practised in line with local conditions, that is to say, different natural resources and farm conditions decide different items of management and animal raising, e.g. various integration of fish - livestock - poultry or integrated management of "aquaculture, agriculture, or a composite of animal husbandry, industry and commerce."

Figure: *Integrated fish farming*

Sericulture

Pond dykes, corner plots, etc. can be used to plant mulberry tree for sericulture on an integrated fish farm. The by-products of sericulture, dregs, which are comprised of the faeces and sloughs of silkworm and mulberry residues can be used as fertilizers for fish culture. On average, we can get 50 – 60 kg silkworm dregs out of 100 kg mulberry leaves after feeding silkworms. If they are used to feed fish, 1,000 kg of silkworm dregs can be converted into 100 kg of fish. By applying 5,000 kg of pond silt to mulberry plots, we can increase 250 kg of mulberry leaves, which can, in turn, raise the production of silkworm cocoon. Pupa occupies 80% of a wilkworm cocoon by weight, and 100 kg of raw silk and 600 kg of pupae can be obtained from every 700 – 800 kg of silkworm cocoon. Silkworm chrysalis is very rich in protein and fat, which are the good feedstuff for fish, livestock and poultry. Pupae have a special ordor, the feeding amount should be controlled. The nutritive components in silkworm drege and pupae are as follows (in percentage):

	moisture	*crude protein*	*crude fat*	*crude fibre*	*non-nitrogen extracts*	*minerals*
faeces	12.2	15.4	2.6	19.6	36.2	
dry pupae	0	46.74	32.22		4.5	4.5

The collection and utilisation of by-products from sericulture:

Silkworm dregs, mulberry residues could be collected at an early stage of silkworm breeding, but the amount is negligible. It is the right time to collect them after the dormant stage. Silkworm pupae are byproducts after reeling. There are two ways to collect above-mentioned by-products:

1. Sundry method: It is suitable for sunny days. Pulverise them after dry, and mix them with chicken feeds for broilers. The daily amount should not exceed 5% of the total feeds. This must be discontinued when the chicken are one month old, otherwise, the meat quality would be affected. The amount of pupae should not be over 10% of pigs' compound fodder. The pupae are allowed to feed sows, piglings, not hogs.
2. Water immersion method: It is for rainy days. As it is difficult to dry silkworm dregs and pupae, just stock these in big vats and fill the vats with water. The preserved ones can be fed to piglings only, never for the fatteners because of the stronger odor of the wet pupae, which may influence the quality of pork.

Chicken Farming

The digestive tract of a chicken is very short, only 6 times of its own body length. Some of their feedstuffs are excreted out before being fully digested.

From the research results, it is found that about 80% of the feedstuffs are utilised and digested by the poultry (count by dry material); Therefore, various animals on the integrated fish farm can use 20% undigested feedstuffs from the chicken manure. Besides, chicken has the habit of pecking foods, more than 10–15 % of the feed are scattered over the ground these can also be reclaimed together with the chicken manure. The total protein contents of the dry chicken is as high as 20–30%. In it is satisfactory if the protein contents of feedstuffs are over 18%. In the past chicken manure used to be fertilizer for the crops which, we consider, is a world of waste.

A little slip will cost a big loss. It is very different from small scale family raising.

The utilisation of good species and hybrid vigor of broilers that are suitable for an integrated fish farm: The most common species of broiler chicken in the world are hybrid offspring of White Cornish and White Rock. Some companies in the world are specifically engaged in chicken breeding. They have got some purebred for the hybrid experiments. From the results of their hybrid matching, they select the best combination of the hybrids. They have first generation hybrids (grandfather generation), and second generation hybrids (parent generation), and then produce broiler chickens for commercial production.

The white feather broilers bred in China are Starbro System from Canada and Hybro System from Holland. We have established National Pure line Chicken Farm, selling grandfather generation to the first grade chicken farms, and the first grade chicken farms sell parent generations to the second grade chicken farms. The second grade chicken farms supplies double hybrid commercial chickens for the markets.

Chinese people prefer to eat yellow feather broiler chickens, therefore, most of the white feather broiler chickens are processed as fronen chicken for export. The live poultry for domestic market and , yellow feather chickens. They are "New Pudong Chicken" bred by the Institute of Rusbandary, Shanghai Academy of Agriculture. The meat of this species is better than Starbro chicken imported, but their growth is a little slower, their body weight reaches 1,500 grams after 10 weeks, a hen can lay 140 eggs per year. Their viability is.

In recent years, some of the developed countries use processed chicken manure to feed animals and poultry. "Chicken Manure Pelletining Production" which becomes the "Regenerative technology Industry" of 80's is of very important economical value. In China, dry saw-dust and pulverised dry stalk of crops are used as bedding material for chicken coops. The bedding material and chicken excreta are used to feed fish. Part of them may be eaten directly by fish as feedstuffs while the rest will become the nutrients of the fish pond water.

In this way, we still cannot fully utilise all the nutrients of chicken manure. For a change, the excreta together with the bedding material are used as the feedstuffs for cattle or pigs, especially for sows and beef cattle.

Thus, the undigested and unabsorbed fine feeds in chicken excreta are well utilised again by sow and cow, promoting the development of the hog and cow husbandry. The excreta of pigs and cattle can also be used to feed fishes. Fishes can utilise the undigested nutrients in pig manure and cow dung. The utilisation rate of feedstuffs is then greatly increased. Besides, animal manures serve as fertilizers in the water to produce abundant plankton, which is a good natural food for fish.

If animal excreta are used to culture fish, they enrich the pond silt, which can be used as fertilizer for the mulberry plot or on the pond dyke for green fodders resulting in high yield.

Thus, we can raise silkworm, also raising chicken, pig, cow and rearing fish at the same time. In this way, the utilisation of the feedstuffs and fertilizers can reach its maximum, and aquaculture and animal raising can be carried out in a more comprehensive way.

Commercial Broiler Production on an Integrated Fish Farm

Characteristics of broiler farming: The broiler farming has only developed for 20 years in China. Large flock and high density farming method is adopted. Broilers grow very quickly. In the feeding condition is favourable they can be sent to market after 8 weeks, usually not over 12 weeks.

The average weight of the chicken is between 1,250–1,600 grams. The feedstuffs of broilers should be compound feed with all the essential nutrients. Special attention must be paid to the temperature, and of the house broiler farming,

Chicken House Construction and Main Equipment Farms From the economical point of view, the open-type house with natural lighting and ventilation for chicken is practiced or fish farms.

Usually, it is a one-storeyed building with inverted "V" shape roof. Its length is about 6–8 meters. Its height from the ground to the caves is about 2.2–2.5 meters even 3 meters in the warmer area. Each room has an area of about 100 square meters, which can accomodate about 800 chickens.

The width of the house should depend on the amount of chickens raised. Two workers can handle about 3,000–4,000 chickens at one time. The chicken house should face the sun, the ratio of the windows and the ground area is usually 1:8–1:10. The windows in the south should be bigger and nearer to the ground while the windows on the north side should be smaller and farther to the ground in the northern hemisphere. It's better to have a concrete ground 30 cm higher than the ground outdoor so that it will be easier to keep it dry and sterilised.

Main Equipment for Chicken Farming

- *Thermafication:* Thermoelectric umbrellas are commonly used. It is square in shape, about 1–1.1 meters on each side, about 0.7 meters in height, with 45 degrees inclination upward. Each umbrella has a 300 W. Thermoelectric wires and a thermostat which can keep 250–300 chicklings warm.

In the cold winter, stoves with pipes should be installed in the house in order to keep the room warmer.

- *Feeding Trough:* There are 2 types of feeding systems, the manuel and the mechanical chain type.

  The feeding trough for chicken is made of wood plate or iron sheet. The size and height depend upon the growth period of chickens. Usually there are small, medium and large sizes.

  There is a special rod on the trough to prevent the spilling of the feedstuffs and the continuation from chicken manure. There should be a certain number of troughs to guarantee the chicken to be fed evenly. This kind of troughs can also hold wet feedstuffs. Cylindric feeding trough can also be used.

  Chain-driven feeding system: It is a trough made of iron sheet. At one end of the trough, attached a feedstuff box, with a chain in the trough moving transversely, which can deliver the feedstuffs to the whole length of the trough. It is easy to use and save manpower.

- *Waterer:* For chicklings a big opening jar with an aluminium plate beneath can be used. They should be installed around the thermoumbrella in order that the chicklings can easily get water. As the chicken grows, automatic barrel-like plastic waterer can be used instead of jar.

  It contains clean water for about 100 chickens to drink per day. Some farms use long trough with running water and the results is also promising.

- *Chicken Cage:* The chick transportation cages are made of calciumplastic corrugated paper, which can hold 100 newly-hatched chickens. For the chickens sent to market, they are usually kept in iron wire cage of 90 × 60 × 35 centimeters in size. Each cage can hold about 15–20 live chickens weighing 1.5 kilograms each.

Standards and Formula of Broiler Chicken Feeds

In order to fully utilise the feedstuffs, to cut down the cost of the feed, to accelerate the growth and build up the health of the broiler chicken, we have to set the standards of the nutrition according to nutritional needs.

The following nutritional requirements are adopted:

***Table:** Nutritional Standards for Broiler Chicken*

	0–5 weeks old	*Over 5 weeks old*
Metabolic energy (cal/kg)	2800–3000	3000–3200
Crude Protein(%)	20–22	18–20
Ratio of Protein and energy (g/kilo cal)	72	61
Calcium (%)	0.9	0.9
Phosphate (%)	0.65	0.65
NaCl (%)	0.37	0.37
Amino Acids (g/kilo cal)	2.66	
Cysteine	1.24	1.09
Lysine	3.91	3.44
Tryptophan	0.72	0.63
Arginine	4.38	3.75
Leucine	5.00	4.38
Isoleucine	2.69	2.34
Phenylalanine	4.68	4.06
Tyrosine	2.18	1.88
Threonine	2.50	2.19
Valine	3.13	2.66
Histidine	1.44	1.25
Glycine and /or Serine	3.59	3.13

***Table:** Nutritional Standards of Vitamins and Minerals for Broiler Chicken*

	Chicken of 0–8 weeks old
Vitamin A Active (I.U./kg)	1,500
Vitamin D_3 (Chicken I.U./kg)	200
Vitamin E (mg/kg)	10
Vitamin K_1 (mg/kg)	.53
Vitamin B_1 (Thiamine HCl) (mg/kg)	.8
Vitamin B_2 (Riboflavin) (mg/kg)	.6
Pantothenic Acid (mg/kg)	10
Nicotinic Acid (mg/kg)	27
Vitamine B_6 (mg/kg)	3
Biotin (mg/kg)	0.09
Choline (mg/kg)	1,300

Contd...

	Chicken of 0–8 weeks old
Folic Acid (mg/kg)	0.55
Vitamin B_{12} (mg/kg)	.009
Sodium (Na) (%)	0.15
Potasium (K) (%)	
Manganese (Mn) (mg/kg)	55
Iodine (I) (mg/kg)	0.35
Magnesium (Mg) (mg/kg)	500
Iron (Fe) (mg/kg)	80
Copper (Cu) (mg/kg)	4
Zinc (Zn) (mg/kg)	50
Selenium (Se) (mg/kg)	0.1

For easy preparation of the feed, there are vitamin compounds and trace elements additives for broiler chickens manufactured in our country. We can just add them to the feed according to a certain proportion.

There are a number of feedstuff companies in the world, they prepare compound feed for broiler chickens. According to the requirements as well as local feed resources, the results are quite promising.

The metabolic energy, crude protein, amino acils, calcium, phosphates, vitamins and mineral contents of the feedstuffs' formula can fulfil the nutritional requirements.

Table: *Average Feeding Standards for Broiler Chicken*

Age by days	Amount of feed (g) each chicken per day	Age by day	Amount of feed (g) each chicken per day
1–5	10.0	31–35	80.0
6–10	20.0	6–40	90.0
11–15	32.0	41–45	100.0
16–20	44.0	46–50	110.0
21–25	58.0	51–55	115.0
26–30	70.0	56–60	100.0

Table: *Formula of the Feedstuffs for broiler Chicken Produces by the Wuxi Feedstuffs Company*

	0–5 weeks old ratio	*above 5 weeks ratio*
Corn	38	48
Highland Barley	15	10
Soya Bean Cake	5	5
Bran Cake	6	8
Cotton-seed Cake	5	5
Wheat Bran	6	5
Low Grade Wheat Flour	5	
Fish meal	9	9
Peptone	4	2.5
Chinese Scholartree *Leave Powder*	4.5	5
Bone Powder	1.5	1.5
Calcium Carbonates	0.5	0.6
Ferrous Sulphate	0.2	0.2
Table Salt	0.1	0.1
Trace element additives	0.2	0.1
Multivitamin additives	5g.	3 g.
Metabolic Energy (Cal./kg)	3036	3057
Crude Protein (%)	21.57	20.14
Crude Celluose (%)	3.71	2.76
Calcium (%)	1.08	1.06
Phosphate (%)	0.86	0.65

Feeding and Management of Broiler Chicken

The most important thing in the Management of broiler chicken is the formula of the feed. The nutrient of the feed should be all-round. Second, the daily management of the poultry should be emphasized. Good environmental conditions are essential to the growth of the chicken. Better rewards could be obtained from the feedstuffs, if we follow the above-montioned criteria.

Simultaneously in-and-out System

All chickens should start to feed at the same day and come to the market at the same day. After all the chickens are sold, complete cleaning and sterilisation of the poultry farm should be carried out. Then rest for 7–14 days. This period of fallowness can break any cycle of any infectious disease. A "clean start" will be ready for the next batch of chickens, which can be prevented from the infectious diseases of the previous batch. As the growth speed of broiler chickens are not uniform, it is better to modify the simultaneously in-and-out system. Chicklings are received at the same time, but sold according to their body weight. For those growing slower, we keep them one to two weeks more, and sell them while their body weight reaches the requirements.

Good Preparatory Work

After all the chickens are sold, the poultry house and equipment should be completely sterilised and a layer of clean and dry bedding is laid on the floor. Before chicklings are coming, the temperature of chicken nursery should be kept above 80°F, and the temperature beneath the nursing umbrella should be 90–95°F, lamps should be installed in the umbrella in order to attract chicklings to come under the umbrella to keep warm.

The trough and waterers should be also installed around the umbrella. Protecting board is set up outside of the umbrella, thus keeping chicklings in the warm area of the umbrella. Waters and trough should be filled before the introduction of chicklings. Chicklings should come to the poultry house in the morning. In this way, they can learn to eat and drink in the day time.

The lamps should be ligned on in the first two weeks Fresh food and water should be given and the daily consumption of the feedstuffs recorded. The temperature of the nursing umbrella should be checked in the night to prevent chicklings from crowding together if it is not warm enough. No light is needed all the night after 48 hours. The protective board of the nursing umbrella should extend and the temperature lowers one degree F daily.

The air should be kept fresh, water and feedstuffs should be clean. The bedding of the chicken house should be dry and the environment quiet. As chicklings grow, the protective board can be removed on the 7th day to avoid too many chickens gathering together. The trough and waterers should also be changed to larger size as the chickens grow. The nursing umbrella should be gradually raised up in height. If feed is given all day long, the feed and water should not be interrupted. If artificial feeding is carried out, the feeding time should be fixed. Weaker chickens should be separated into groups for feeding. The bedding on the ground should be changed frequently and exposed to the sun as often as possible.

The chicken house should be kept warm in winter and well ventilated and proper measures should be taken to lower the temperature in summer.

Chicklings are fed with special diet during the first 5 weeks and then fed with fattening diet gradually. Prevention of diseases should be emphasized during the whole course of raising. When those chickens reach 10 weeks old, with body weight of 1.5 kg, they can be sold first. It is better to catch the chickens in the early morning and each cage can hold 15 chickens in the summer and 20 chickens in the winter. A layer of dry grass is laid on the base of the cage. Tender can during the catching can avoid injuries.

Prevention and Treatment of Common Disease

The raising time of broiler chickens is short, but the density is high. The prevention of disease should stress nursing management, cleanness and hygiene. The prevention program should be outlined ahead of raising. Medical treatment is carried out only hen the prevention fails.

The common chicken diseases are Pullorum disease, coccidiosis avium, infectious bronchitis of chicken, chronic respiratory diseases, variola avium, Newcastle disease, cholera avium, Marek's disease, ascariasis avium and deficiency disease, e.g. lack of certain nutrients, such as vitamins and minerals.

Some of the chicken diseases can be prevented by inoculation of vaccine or by medication. Some big manufacturers have produced mixed whole feedstuffs including vitamins, trace elements and drugs, thus, it can effectively prevent the occurrence of certain diseases. Some of drugs may produce resistance after long application, such as anti-coccidiosis medication.

Therefore, the most helpful way to prevent the outbreak of chicken diseases is good nursing management, cleanness and hygeine as well as isolation and sterilisation.

Example of Preventive Project

Marek's vaccine is given to chicklings within 48 hours after hatching. 0.01% furazolidonum is added into the drinking water for 1–7 dayold chicklings to prevent Pullorum disease.

Nose drops of weak toxic Newcastle disease II vaccine are given to the 7 day-old chickens, and they are inoculated with variola avium vaccine at the same time. Robenidine of 30 ppm is added into the feed after 10 days old to prevent coccidiosis and discontinued one week before marketing.

The beak should be cut in the middle with a beak cutter and hemostasic is done by cautery at 15 days old to prevent bad pecking habit.

H_{120} vaccine of infectious bronchitis is given with the drinking water to 20 day-old chickens at the dilution of 1:500–1000. The dosage is 5–10 ml per chicken.

Subcutaneous inoculation or Newcastle disease vaccine at 45 days old. If ascarid is found, tetrametrasol is given.

Control of coccidiosis: Coccidiosis is the most serious disease of broiler chickens and usually happens after the chickens are 8 weeks old. It is mainly transmitted through manure, especially in the hot and humid season.

If the bedding material is thick, the transmission will be more serious because the wet bedding increases the chance of hatching of occyst.

The most effective approach for preventing coccidiosis is to raise chicklings in cages above the ground. Thus, the chickens are isolated from their own excrement.

If thick grass bedding is used, the most common approach is some anti-coccidiosic medication given with the feedstuffs right after initial feeding, and should be carried out through the whole period of growth. For instance, 125 ppm of sulfadimethoxine (SDM) or 30 ppm of chlorophenyl-quanide can be added into the feedstuffs. Dinitolmide (Zoalene) can also be added into the feed, up to the concentration of 0.0125%.

Attention should be paid to that continuous application of anti-coccidiosis drugs for a period of time may bring about the drug-fast and drug-fast strain. As a result, coccidiosis could break out again then, the treatment of coccidiosis is only one of the approaches in the whole process of prevention.

If the result of the first trial is not promising, other medications should be considered. Two kinds of medications can be applied alternatively.

Besides the medical control of coccidiosis, cleanness and hygiene of the chicken house and the dryness of the ground bedding are also essential. The addition of vitamin A and K in the feed can increase the resistance of the chicken.

Collection and Utilisation of Chicken Manure

***Table** : Nutrient Constituents in Chicken Manure (Dry)*

	Cage Raising	*Ground Raising with saw dust bedding*	*Ground Raising with dry grass bedding*
Moisture (%)	11.4	12.3	15.5
Crude Protein (%)	26.7	21.0	
Crude Fat(%)	1.76	1.7	2.3
Non-nitrogen Extracts (%)	30.6	30	27.1
Crude Cellulose (%)	13.04	17.2	18.7
Minerals (%)	16.9	16.9	14.1
Calcium (%)	7.8	1.95	2.3
Phosphate (%)	2.2	1.26	0.42

Deodourising of Chicken Manure

The chicken manure is fermented and decomposed a few hours after excretion, and a bad smell will occur. Animals instinctively refuse to eat chicken manure unless they are accustomed to it. It is better to deodourise the chicken manure before feeding.

Adding Ferrous Sulphate: Ferrous sulphate for industrial use ($FeSO_4$ $7H_2O$, green vitriol) has a mild bactericidal function. After mixing it with chicken manure, there will be no fermentation, no decomposition and no foul smell. But ferrous sulphate will absorb moisture, which makes the pulverisation and spreading of the powder difficult. Therefore, the process of defouling is limited. Half amount of chimney dust can be added and mixed, then baked dry at 60–80°C, the mixture will not absorb moisture, and this makes the spreading easier. 7% of ferrous sulphate and 3.5% pulverised coal dust are blended with chicken manure. After drying, it will become odorless chicken manure, which can be tolerated by pig, cow, fish, even chicken itself. Their nutritional value are very high, and can be used up to 20% in the feedstuffs.

Moisture	*Protein*	*Fat*	*Non- Nitrogen Extracts*	*Crude Cellulose*	*Minerals*	*Caloric*
%	%	%	%	%	%	cal/kg
10.35	21.9	1.25	21.33	7.13	30.04	2161

Fermentation Method: If dried grasses are used as bedding, the chicken manure and the dried and pulverised grass (about 14%) are put together into a fermontation pool with dimensions of 2 × 2 × 0.5 m and then, add in equal amount of dried and pulverised grass, silage, wheat bran and a small amount of table salt. The mixture will be filled with water, which occupies 70% and be sincked to 15 cm in height, fermenting for 4 hours in summer. If the temperature is over 40 degrees C, spread out the fermented material. It should be heaped up to 20–30 cm in the spring and fall, let it ferment for about a day, and it should be stacked up to 40–50 cm in height in winter, and covered with a plastic sheet for about 3–4 days. This kind of formented feed has no foul smell at all, but with some wine flavour, good for taste, and pigs like it very much.

Utilisation of Fresh Chicken Manure: The nutritional value of the manure of less than 5-week-old chicken are very high; pigs like it even without any treatment. From the experiments, the manure of the egg-laying hens contains about 60% of the crude protein in the feedstuffs. There are 4 grams of digestible protein in the excreta of the egg-laying hen per day. The fresh chicken manure from the cage-cultured hens can be collected by the manure collector and mixed with the feed immediately for the pigs feed. If pigs are accustomed to such a kind of food, they can be utilised more efficiently. It is better to feed sows, and should not be fed to fatteners, which may influence the quality of pork.

Sterilisation of Chicken Manure: The chicken manure collected contains latent pathogens, sterilisation is needed so as to prevent other animals from chicken diseases. It is considered to be the most appropriate approach in the foreign countries to use methyl alcohol bromide (MeBr) to fumigate manures including bedding. Fumigation can be carried out in the storage basement of silage, or in the special designed tanks.

Another method is to stir up the manure with formalin solution at a dose of 0.5–2% and to put it into use after air-dry.

Chicken manure fed to cow and pig should be limited within 10–30%. If 15–25%, the results will be promising.

Goose Production

Both goose and duck are water fowl. Goose raising is mainly for human consumption. Goose is herbivorous animal, which can utilise some of the green scarse feeds. Not like ducks, geese don't have to use feeds of animal origin. Therefore the cost of the feeds for geese is much lower than that for ducks. The well known Chinese goose has strong ability to graze with early maturation and high output. The ratio of the feeds and meat, that is the reward of feed, is higher for the goose than for the duck. It is more economical to raise goose.

Fishpond can also be used to raise goose. The goose droppings go directly into the fishpond, which can fertilize the pond water and can be fed to fish as well. If ducks are raised in the fishpond, they will take small fishes as their food. With geese in the pond is a good natural scenery, "white feathers float on green water; Red palms pluck blue ripples."

Taihu Lake Goose is a Fine Species for an Integrated Fish Farm

Chinese goose is famous for its early maturity and high output. They can be divided into large and small body-types according to their size. Lion-head goose is the largest variety of Chinese goose, and also one of the largest varieties in the world. Matured goose weighs about 10–12 kilograms, gosling grows very quickly, 75–90 day-old meat goose can weigh about 5–7.5 kilograms. But their ability to seek food is poor and they need more feedstuffs, bringing forth only 25–35 eggs per year. Taihu goose is famous for small body-type white goose in China, its size is not big, the matured goose only weighs about 3.5–4.5 kilograms. But they mature earlier and can lay about 80 eggs per year.

High yield species can lay more than 100 eggs per year. They have a good ability to seek food and the consumption of foodstuffs is less; therefore, Taihu goose is better than others from the economical point of view. We take small size, early matured, high yield Taihu goose as maternal-line and large size lionhead goose as paternal-line, through artificial insemination, cross matching and heterosis to produce meat gosling.

Grazing of Meat Gosling

Spring is the grazing season for the meat goslings. The weather in spring gradually becomes warm and grasses sprout. It's good time for the growth and grazing of the goslings. After wheat harvest, goslings can graze in the wheat field and they can seek the remaining wheats to fatten themselves, therefore, less feedstuffs are required. Only about one kilogram of feedstuffs is required for each gosling up to 70 days old.

The newly -born goslings are very timid and afraid of cold and like to get together, therefore, they should be kept warmer and divided into small groups. They have to be fed for at least six times per day at a fixed time, and midnight feeding is essential. Feedstuffs can be fragmentary rices, chopped green cabbages or green grass. After 4 or 5 days, they are allowed to graze and the distance of the grazing can increase gradually. After 15 days they can go out for camping. They should be kept in the hut for the night to prevent injuries from the wild animals. During the fattening stage, if the weather is hot, they are kept in the rivers, lakes and ponds for the night. Up to 70 days, the average weight reaches 2.5 kilograms, and then they can be marketed.

Management and Feeding of Breed Goose

Goose is a water fowl. Integrated fish farm can use fishpond to raise goose. A simple goose house can be erected on the bank of the fishpond. The goosehouse should face the south, the southern side is open or enclosed by low walls. There should be a dry run of at least 5 meters in width for feeding and resting. A slope of less than 30 degrees is to connect the dry run with the fish pond.

The fishpond is enclosed with hedges about 60 centimeters in height. The ground of the goose-house is 20 centimeters higher than the outside and grass is used as bedding to keep the house dry. In one corner of the house, straws are stacked for laying eggs. Each square meter of the house can hold about 4 Taihu geese, but the water area should be as large as possible. In order to get high fertilization rate of seed egg, there should be 15 male geese matched with 100 female geese. Taihu goose can utilise green fodders; therefore, during non-egg laying season, raising breed geese depends on grazing with a small amount of blighted rices as supplementary food.

The feedstuffs should be increased upon the egg-laying stage. With the increasing production of the eggs, the ratio of fine feeds should increase in proportion and special attention should be paid to the green feeds and minerals supplies. In order to promote the egg-laying of the breed geese, lamps should be put on in the night so as to prolong the lighting time from 12 hours to 15 hours. Grazing should be performed as much as possible. The feedstuffs ingradient and amount of feeding of the Taihu goose would be as follows:

Table: *The ratio of Feeds for the Breed Goose*

Month	*Daily feeding amount*	*feed ingradient (%)*					*Metabolic energy (Cal/kg)*	*crude protein (%)*
		blighted rice	*bran*	*bran cake*	*Frag rice*	*barnyard grass*		
Jun-Aug	150	50	50				557	3.78
Sept-Nov	250	50	20	15	15		842	7.34
Dec	250	40	20	15	15	10	1346	9.75
Jan-Mar	200	40		20	20	20	1746	9.75
Apr-Jun	175	30		20	20	30	1748	10.28

In the above-mentioned ratio of the feedstuffs, the nutrients, especially the crude protein and metabolic energy are too low, but it is due to limitation of the food source and raising cost, this is the best ratio we can obtain. As Taihu geese have good propensity to graze and good ability to utilise coarse feed, they can adjust their intake according to the requirements of the energy and protein. Thus, they can overcome the disadvantages in the formula. If we can improve the formula of feeding, the egg-laying rate can be raised.

Common Goose Diseases

Matured geese are rather strong and have good resistance against diseases and do not easily get sick. But the resistance of the goslings are very weak.

If they are kept in large goosery with dense population, any little slip in the management, that is against the physiological requirements of the goslings will cause death. The best preventive approaches to the goose disease are good rearing management, sanitary environment and proper vaccination

The common infectious diseases among the goosery are gosling pest, cholera avium and yolk peritonitis. Gosling pest vaccine is now available in China to prevent its occurrence. The mother goose should be inoculated with diluted 1:100 gosling pest vaccine one month before egg laying, once a year, and then the goslings hatched will have a strong immunity. Regular inoculation of cholera avium vaccine can prevent cholera of the geese. If the goosery is already infected, streptomycin should be injected at a dose of 100,000 IU for each goose at four-hour interval for three times, and the result is promising. Yolk peritonitis is usually occuring during the egg-laying season and there is no such vaccine for prevention now. However, streptomycin can be used for treatment, and the whole goosery should be fed with furazolidonum mixed with the feeds at a dose of 25 mg for each goose for 3–4 consecutive days. No matter what communicable disease is, besides medical control, other measures such as sterilisation, isolation and deep bury of carcase ought to be taken timely to prevent the spreading of disease.

Duck Production

Chinese people have a tradition of eating duck eggs and processing them into salted or preserved eggs. Raising of egg-laying duck needs less and simpler equipment. They mature earlier, produce more eggs, and the size of the egg is rather large.

Ducks live in groups, and feed on natural feeds through grazing. Therefore, the raising of egg-laying duck is more profitable than that of egg-laying hen. On an integrated fish farm, grow-out pond can be used to raise egg-laying duck. Duck is an omnivorous water fowl which can use the by-products of the crops and animal feeds as well. Duck manure has higher nitrogen contents, which doubles that of goose manure, and the phosphate contents are nearly twice that of goose manure.

1. Shaoxing duck is one of the good egg-laying species in China: Its somatotype is rather small, the average weight of the matured duck is about 1.25– 1.5 kilograms. Its ability to seek food is rather strong, so it needs less artificial feed. It matures earlier and usually lays eggs at the age of 120 days. 250–300 eggs can be produced every year. High productive groups can give more than 300 eggs/duck per annum. Each egg weights about 55–65 grams. We have introduced drake of Khaki-Compbell strains from Holland and crossed them with Shaoxing female duck. The hybrid vigor is rather obvious: the first generation hybrids start to lay eggs after 102 days, averagely producing 287.7 eggs per year. Each egg weighs about 64.5 grams. It can produce about 18.8 kilograms of eggs per annum. The economic benefit is about 50% higher than that of Shaoxing duck.
2. Maximum utilisation of natural resources for grazing: Duck is an omnivorous water fowl. Their ability to seek food is very great; therefore, they should be raised by grazing as much as possible. We can utilise the natural resources in the rice fields, lakes and marshes, trenches and ditches or river shoals. Egg-laying ducks usually hatch their ducklings in the early fall, the paddy field can be used for grazing. Ducklings eat grass, snails, small fishes and shrimps and injurious insects in the paddy field. After the rice grows up, transfer the ducks into the rivers, channels, ditches, lakes or marshes for grazing. After the harvest of rice crops, duck can be driven back to rice paddy fields where there must be some left-over rice, barnyard grass, snails and mole crickets. The natural feeds are seasonally available, duck raising must take advantage of natural resources so as to save artificial feeds. When grazing stops by November, ducks start to lay eggs.
3. Raising Ducks in Fishponds: Fishponds on an integrated fish farm can be used to raise ducks, but ducks eat small fishes,

and they dive into water competing with black carp for snails. Therefore, we had better raise the ducks in grow-out ponds with much bigger water area. The construction of duck coops is similar to the construction of goose house, but egg-laying ducks are more sensitive to cold in comparison with geese. Lower temperature will affect the egg laying, so duck coops should be kept warm, there should be windows and doors on the southern wall to keep coops cool in summer and warm in winter.

If there are some lakes, marshes, river shoals with plenty of natural feeds nearby the duck farm, grazing should be utilised to its utmost. Otherwise, they could be bred in suck shed, moving around in dryrun, wetrun and fishpond for food and rest, and all the egg-laying ducks will live mainly on artificial feeds.

Each Shaoxing duck has to be fed with 110–120 grams of mixed feedstuffs, including blighted rices, barnyard grass, rice bran and fragmentary rices, and 50 grams of animal feeds, such as fresh fishes and pupae. Besides 100–200 grams of green vegetables for each duck per day, snails and Corbicula spp. are often to be supplemented to meet their calcium needs.

The green fodder should be chopped and fresh fishes should be cooked and then be mixed with green fodder. Feeds are given three times daily at set time. Shaoxing duck is rather nervous, easily to be disturbed, which may influence their egg-laying ability. Therefore, we should keep the environment very peaceful and quiet. Although egg-laying duck is a water fowl, they have to rest on land and thus, the duck house should be dry and grass bedding should be thick in winter in order to keep it warm and dry.

Prevention and Treatment of Common Diseases

The most important infectious diseases of egg-laying duck are duck pest and fowl cholera. We now have vaccine for preventive inoculation against the pest.

The results are very promising, while the immunological effects of the fowl cholera is still unsatisfactory. To treat fowl cholera, each egg-laying duck is injected intramuscularly with 50,000 IU of streptomycin at an interval of 4 hours for three consecutive times, then followed by oral ingestion of 0.2 grams of sulfadimethoxinum daily for 3 consecutive days. Sterilisation, isolation and deep bury of carcase should be taken to prevent the spreading of the disease.

Pig Farming

Pig is an omnivorous animal, the length of pig's digestive tracts is 14 times of its body length. Feedstuffs are much better utilised by pig than by chicken. Chinese pigs can tolerate and fully utilise coarse fodder, even chicken excrement including the bedding. After being accustomed to it, pigs can take a large amount of these fodders. Pig has become one of the main animals in the complete set of "fish-livestock-poultry" system on an integrated fish farm.

Pigsty on the Pond Dyke

Pigsties usually are built on pond dyke on an integrated fish farm, so that the excreta of pigs can be directly flushed into the pond. Because the pond dyke is not too wide, the pigsty is usually of single row, facing the south with inverted "V" -shaped roof.

The depth of the pigsty from south to north is 4 meters, the width of every house 3 meters. There is a veranda about one meter in width on the southern side. Each house has an area of about 11 square meters and can hold 10 fattening hogs. The height from the ground to the eaves can be 2–2.2 meters. In order to reduce the expenses, no wall is needed on the southern side, but only a fence about 1.2 meters in height. The cement ground is a little higher in the northern side, and slightly sloping down toward the south.

Thus, pigs' excreta can easily pass out directly into the fishpond through discharge ditches. This kind semi-open type pigsty has the advantage of less building cost, good ventilation, sufficient sun-light, cooler in summer and warmer in winter. The feeding trough is installed on the southern side and pigs sleep on the northern side. Thus, the ground can be kept dry and it is convenient for operation and management.

Utilisation of Hybrid Vigor in Pig Farming

Taihu pig bred in the Wuxi district can copulate when it reaches 30 kilograms in weight at 5 months old. Each litter has about 15.5 piglets on average. Matured sow weighs about 125 kilograms. They mature earlier and give more piglets for each litter.

The quality of the pork is good, less fodders are used and they can tolerate coarse fodder, but the growth rate for the fattening pig is not satisfactory. We use yorkshire stud pig for matching and use their first generation hybrid to produce fattening hogs with a good result.

As people's living standard has been improving, more lean meat is required. Landrace hog for salted meat is being recommended for cross matching. In the district where fodders are abundant, the first generation hybrid sow from the Yorkshire stud pig and Taihu sow is intercrossed with Landrace stud pig. The offsprings of this kind of triple hybrids grow very rapidly, and their hybrid vigor is more evident. In order to fully utilise the good stock, artificial insemination is often adopted for matching.

Fodders of the Pig and Rearing Management

On an integrated fish farm, there are various kinds of fodders which can be used as pig feeds, but we should have reasonable arrangements in order to increase their nutritional value, and promote the growth and fattening of the hogs.

In order to lower the cost of pig farming, local produce and by-products of the farm should be utilised to the maximum. The feeds should be complete in nutrition, including 55–60% of corn, barley, wheat bran and rice bran, 5% of soya bean cake and cotton-seed cake, 15–25% of chicken manure, 15–20% of wine lees and bean dregs, with an addition of 0.5% of table salt and 2–3% bone power and calcium carbonates. Green fodder is about 15–30% of the total fodder.

The piglets may be the ones reproduced from the sow of the same farm, but usually the piglets are bought from the breeding farm. Piglet fodder should be given to the newly bought piglings and young pig-fodder given after the body weight reaches over 20 kilograms. When the body weight reaches 30 kilograms or more, fattening fodders are maintained until the body weight is over 90 kilograms.

The following formula of the feed can be used in the area where is rich in fodder crops;

Weight of pig (kg)	*Piglet's fodder 5–20*	*Young pig's fodder 20–50*	*Fattening fodder 50–90*
Item			
corn (%)	53	50.5	45.66
Barley (%)	7.5	15	11
Sorghum (%)	12	5	11
Soya bean cake (%)	15	12	7
Wheat bran (%)	-	5	15
Fish meal (%)	10	5	3
Chinese scholartree			

Contd...

Weight of pig (kg)	*Piglet's fodder 5–20*	*Young pig's fodder 20–50*	*Fattening fodder 50–90*
leave powder (%)	-	5	5
Bone powder (%)	2	2	2
Table salts (%)	0.5	0.5	0.35
Total (%)	100	100	100
Additive to each ton of fodder			
Sodium selenite (g)	0.15	0.15	0.15
Zinc sulphate (g)	200	200	200
Potassium iodide (g)	1	1	1
Multivitamins (g)	40	40	40
Digestive energy (Cal/kg)	3,084	3,011	3,028
Crude protein (%)	19.3	16.41	14.32
Daily amount for each pig (kg)	0.3–1.2	2.1–2.2	2.2–3

The piglets should be trained right upon arrival in the pigsty. They are trained to eat, sleep and discharge at a fixed place. The feeding trough is usually placed on the southern side of the hogpen, and a small amount of manure is put near the corner of the manure exit on the southern side in order that the newly-arrived piglets smell the excreta, and then have a habit of discharging at that place.

Clean grass bedding is stacked on higher place on the northern side for sleeping. Sometimes, piglets discharge everywhere in the hogpen, the stockman should sweep the excreta into a fixed corner to keep the sleeping place clean. After several days, piglings could form a good habit and keep their sleeping place dry and clean. The piglets are fed three times a day with a fixed feeding time and amount. Hoghouse needs to be cleaned 2–3 times every day to make the piglings sleep and eat well.

Prevention and Treatment of Common Disease

The resistance of pig is stronger than that of chicken. If we can take good care of their health, including proper management, sanitary environment and preventive measure, pigs are not easy to get sick.

The common infectious diseases of pigs are: pig pest, pig erysipelas, pig pasteurellosis, piglet parathyphoid, and pig astheme. There are vaccines for the former four diseases, and "triple vaccine" for these three fulmingating infectious diseases, pig pest, pig erysipelas and pig pasteurellosis. It is convenient to use and very effective.

There has been no vaccine for pig asthema so far. The main measures are quarantine, eliminating diseased pigs and raising healthy pigs. Intramuscular injection of kanamycin, 20,000– 40,000 IU/kg of body weight once a day for 5 consecutive days or intramuscular injection of oxytetracycline, 20–40 mg/kg of body weight once a day for 5–7 consecutive days will have certain curative effects. The sick pig should be isolated, sterilised and dead pigs should be deeply buried or burned in order to stop the spreading of the disease.

Milk Cow Production

Cow is a herbivorous ruminant animal, and it can better utilise green forage, especially the chicken manure including grass bedding. Chicken manure contains a large amount of non-protein nitrogen compounds which exist in a state of uric acid or amide and can not be effectively absorbed by the pigs, but milk cow as a ruminant animal, can well utilise them. That's why milk cow is one of animals raised in "fish-livestock-and-poultry" integration. The milk output of the cow is high, economic benefit great, and excreta considerable. The main approach of raising milk cow is grazing. So a pasture is needed and stall feeding is practised in winter. Because of lack of large pasture, intensive stall feeding is adopted for raising milk cow on fish farms. In front of double-rowed cowbarn, a play ground is enclosed for cows. They are fed and milked in the stall, and walk around in the play ground. All the forages needed by the milk cow are supplied during stall breeding.

Black and White Cows Suitable for Fish Farm to Raise

Black and white cow is the highest milk yielding species. Most of the countries in the world raise this kind of milk cow, which originated in the Netherland, but they all had bred their own species named after their own countries such as, American black and white cow, Japanese black and white cow etc. Chinese black and white cow is one of the main species of our milk cows which can acclimate the natural conditions of the local district, and with high productivity of milk. The primiparous cow could yield 4,000 kilograms of milk during 305 days of lactation, and over 5,000 kilograms of milk can be expected after third pregnancies. The milk fat index is 3.2– 3.5%. The Chinese black and white cow has high productivity, strong body and good adaptation, therefore, it is a good species for raising.

Establishing Sound Cow Groups

Tuberculosis and brucellosis are common infectious diseases for both man and cow.

The characteristics of tuberculosis is to cause caseous degeneration of the tubercle nodes on the lung and lymph nodes while brucellosis is to cause abortion, infertility and decrease of lactation of cow.

These two diseases will influence both milk cow and human being, therefore, strict quaratine : measures should be taken in case a fish farm wants to develop dairy enterprise.

Newly-bought milk cow should be isolated for 3 months and tuberculin tests should be performed thrice, by both intradermal inoculation and intraocular instillation. They can not mix with other cows until we are sure that they are disease free. Quarantine test should be performed twice yearly for those cows without tuberculosis.

If there is any cow getting tuberculosis, it should be either isolated immediately or checked out. In that case, all the other cows should be examined again 30–45 days later until there is no positive case found for thrice.

All the workers in the dairy should be regularly examined by X-ray. If any tuberculosis case is found, the patient should be transferred out from the dairy.

If calf is born from a diseased cow, it should be fed separately. Enhance the breeding of healthy calf to replace diseased cow and thus, new healthy cow group can be established. Serum agglutination test can be used for the quarantine control of brucellosis of milk cow. It should be performed yearly.

Healthy cow uses sheep type No. 5 brucellus bacillus (or pig type No. 2) attenuated vaccine for prevention, but they have some bad effect on the quarantine test.

Forage and Feeding of Milk Cow

Stall feeding is normally carried out in the dairy on fish farms. The feeding of milk cow should be performed on the scientific basis; therefore, the formula of daily feeds and feeding plan must be worked out according to the standard which includes nutrient components of the forage and nutritional requirements of the animal.

These standards vary in different countries, each has its own general standard. For instance, U.S.A. has NRC system.

Feeding Standard for Milk Cow (Daily)

***Table:** Oat Unit Standard*

Body weight (kg)	*Oat forage unit (kg)*	*Digestible crude protein (kg)*	*Ca (g)*	*P (g)*
350	3.7	210	18	9
400	4.0	230	20	10
450	4.2	240	23	12
500	4.6	260	25	13
550	4.9	280	28	14
600	5.1	300	30	15
650	5.4	310	33	17

***Table:** Milk Net Energy Unit Standard (NND)*

Body weight (kg)	*Dry materials (kg)*	*NND*	*Digestible crude protein (g)*	*Ca (g)*	*P (g)*	*Carotene (g)*	*Vitamin A (I.U.)*
350	5.04	9.17	227	21	16	37	15
400	5.57	10.13	250	24	18	42	17
450	6.09	11.07	274	27	20	48	19
500	6.58	11.97	296	30	22	53	21
550	7.08	12.88	318	33	24	53	23
600	7.55	13.73	339	36	27	64	26
650	8.02	14.59	360	38	30	69	28

These are the minimum standards, and 10–15% can be added in practice. As the body weight of the primiparous lactating cow is still increasing, their standard should be 20% higher.

If the proposed milk production is over 6,000 kilograms, usually they are fed with the feed for the weaned multiparous cow with daily milk production of 15 kilograms.

Daily portion of forage for the milk cow is given according to the milk production (the amount of milk produced and its fat content index). The feeding standard can be calculated out according to the nutritional requirements of daily milk production plus the average body weight of the milk cow (counted as 600 kilograms). Then take the nutritional requirements of the feeding standard for milk cow for reference, calculate out the total daily nutritional requirements.

The physiological characteristics of different stages of the milk cow should also be considered in the calculation, e.g., pregnancy, etc. The price of the feeds and palatableness should also be considered. The forage of the milk cow is mainly green and coarse feeds. In winter or hay season, cows are fed on one kilogram of dry grass and stalk of the crops, 3–4 kilograms of silage or 5–7 kilograms of root tubers for every 100 kilograms of body weight. 8–10 kilograms of green grass are given to every 100 kilograms of body weight in grass growing season. One kilograms of mixed fine forage is given for production of every 2.5–3 kilograms of milk.

The ratio of fine forage mixture is 30% cake type feedstuffs (if chicken manure are used, only 15% can be added), grain feedstuffs (barley, corn) about 40%, bran 10–15%, by-products of processing 10%, minerals, salts and fish meal 10%. The above ratio is only for reference. The nutritional components of different forages can be calculated from the nutrients table, and also the amount should be adjusted according to the needs of the milk cow. The production of mixed forage developed very quickly in the recent two decades. Forage manufacturers use computors to calculate the needs of the milk cow, and mix the feeds automatically. They can produce and supply mixed whole forage.

Milk cow is fed three times daily with fixed amount at fixed time, adopting the method of small amount but frequent supply during feeding, usually fine forage given first, followed by fodders, and then water. The forage must be fresh. Iron nails and wires must be removed if there is any in it. Milk production can increase by 10–15% if sufficient amount of water is drunk, by milk cow. They should have sufficient exercises. The whole body of the cow is brushed before milking, dry brushing in winter, and washing and brushing in summer. Cow shed should be cleansed frequently in order to keep the shed and playground clean and sanitary. Cow manure and urine are washed directly through manure ditches into fishponds. 5–20% bleaching powder emulsion or 5% cresol is used for the sterilisation, which should be performed at least monthly.

Within 4–5 days after parturition, the milk in the breast of the high yielding milk cow should not all be squeezed out, only 2 kg of milk each time. One third of milk is milked out on the second day and then the amount could gradually be increased. All milk could be squeezed out on the 4th day. Dry fine grass, not forage with much juice and fine food, are given to the weak cow as its main food within 3 days after delivery. The milk production gradually increases 10–15 days after delivery, and reaches its peak of lactation. Fine quality grass, sufficient forage and sufficient amount of water should be guaranteed at this stage to meet the nutritional demands of lactation. The milk production gradually decreases 3 months after delivery, and weaning starts 60 days before next delivery. 10 days before weaning, reduce the amount of fine feeds, green fodders and juicy forage and times of milkings as well. Do milking every other day or every 3–4 days. Stop milking when the production drops to 4–5 kilograms.

Milking Technique and Storage of Fresh Milk

Milking can be conducted manually or electrically. Integrated fish farms usually don't have large numbers of milk cow, therefore, manual milking is preferred. Wash the udder with 50°C warm water and then thoroughly massage the udder. Milking is performed by first massage of two nipples at a time. The first and second runs of milk are collected in a special container, and should not be mixed with normal milk in the milk pail. When a large part of the milk is milked out, the udder should be massaged again in all areas, and finally massage should be carried out the third time at the end of milking until all the milk is out.

The milk collected by manual milking should be filtered by gauze to remove hairs, dust, fecal materials and other impurities. It is to be cooled quickly and preserved in a cool place. The milk should be thoroughly sterilised and covered tightly and transported to milk-collecting station as quick as possible.

Artificial Breeding and Utilisation of Earthworm

Earthworm is good food for fowl and fish. Fresh earthworm contains 8–10% protein, while dry earthworm up to 56–66%. Its effective energy is 2920 Cal. per kg. Its nutritional value is equal to that of fish meal.

The reproductive ability of earthworm is very strong, and it can multiply 200 times within one year under normal conditions. If good care is taken it can multiply 1000 times. No special equipment and fodders are required for the breeding. The fermented cow dung, pig manure, weeds and rank grass from the integrated fish farm mixed with proper amount of silt are good fodders for the earthworm.

Large-scale propagation of earthworm can not only improve the soil, but also increase the productivity of the crops and improve the environment. Artificial breeding of earthworm is now advocated and propagated in our country. Some of integrated fish farms take earthworm breeding as one of their animal raising. Earthworm can serve as protein feeds for both poultry and pigs or fish. The results are very promising in enhancing yields.

Plant Cultivation on an Integrated Fish Farm

Fertilizers and feeds are considered to be two of the basic conditions to strive for a high yield of fish. All fertilizers are used primarily to propagate natural organisms for fish to eat. Various kinds of forage grasses, beans, grains, melons and vegetables, aquatic plants, etc. are also good foods for fish. To establish an integrated fish farm needs certain amount of forage land and land for forestry, animal husbandry and side-line production as well, which are all directly beneficial to fish farming.

Crop production on an integrated fish farm means the utilisation of pond dykes, river banks, corner areas, etc. and the forage field. Cultivation of grains, beans, pasture grasses, melons and vegetables, fruit and mulberries should be rationally practised in accordance with the needs of fish and the growing seasons of different crops. Furthermore, water surface at the edge of rivers & lakes in the vicinity of the fish farm can be utilised to cultivate aquatic plants so that high quality forage will be available at every season of the year. This kind of forage and silage can directly be used not only for breeding fish, but for rearing domestic fowls and animals or fertilizing water to grow plankton for fish.

Pasture Grasses

Perennial Ryograss (Lolium Pereme): It is a perennial plant belonging to the Gramineae family. It has such features as quick growth, high yield, rich in nutrition, easy to cultivate, low cost and strong adaptability. Its yield may reach 5–10 tons per mu. It is a good food for Grass carp, Chinese bream and Wuchang fish. Ryegrass cultivation can provide fresh food for fish in early spring, thereby, enabling them to break their fast ahead of season.

Seeding and Transplanting

Ryegrass is a tardy plant which grows even in the shade and likes moisture. Seeding is usually done at the end of September. Land should be prepared by applying 1,000–1,250 kg/mu of human feces as the base manure, or splash a layer of river silt over it, and upturn the soil by deep tilling, and then pulverise and smooth the soil. The amount of seed for broadcast sowing is 2–2.5 kg/mu. If the weather is dry after sowing, sprinkling is necessary and this should be continued till the complete emergence of seedlings. Then give another dressing of human feces of the same amount as the first application.

Transplantation is usually conducted at the end of October and early November. When seedlings grow to a height of 12–15 cm, they are pulled out to be transplanted in a field that has already been tilled and splashed with a layer of pond silt as base fertiliser. The leaves of seedlings should be cut half to ensure quick establishment. The spacing is 18 × 18 cm or 21 × 21 cm, and 6–7 seedlings are planted in one bunch. After transplanting, apply night soil in 1:3 dilution (night soil 1, water 3) to facilitate establishment and growth.

Field Management

Weeding: In order to enhance the growth of transplants, all weeds have to be eradicated. Weeding should be repeated now and then after transplanting.

Watering: Seedling plots and transplant fields should be kept moist and watering is necessary in prolonged fine weather.

Fertilizer Treatment: Usually one application of nitrogenous fertilizers is given before the initial cropping. Later on, organic manures should be dressed at a rate of 250–300 kg/mu after every harvest. The result is better if soil could be loosened and fertilizer applied the very day when it is mowed. Some inorganic fertilizers can also be used.

Harvesting

When ryegrass grows to a height of 30–60cm, cut them near the ground to serve as fish feeds. At this time, this grass is tender and so every bit of it is consumed.

Thus, the utilisation rate of feed is high. Besides, ryegrass grows fast and has a high power of tillering after cutting. Generally speaking, when air temperature is low from October to next Feb., the growth of ryegrass is retarded, and only one or two outtings could be performed. From March to May, harvesting can be done once every 20 days or so. Be sure that the second and third cuttings are made closely to the ground, leaving stubbles only 2–3 cm high so as to increase tillering, harvesting and promote its quality.

Reservation of Seeds

English ryegrass begins to shoot, bear ears, and flower in April, and seeds can be collected in early June. Since seeds do not ripen at the same time and will fall to the ground easily, so when the colour of the ears turns yellow, collection should be done in time to prevent seed loss. The yield is about 50 kg/mu. Plots reserved for seed production should not be cropped as green fodder.

Integrated Fish Farm: Design and Construction

Aquaculture Engineering

Design and construction for aquaculture facilities: Aquaculture draws on well-established engineering fields for most of the design and construction needs of its production facilities. Building earth ponds is similar to building roads — a knowledge of the characteristics of soils and the limits of safe design are the basis of good construction. Similarly, the buildings used for hatcheries and other support activities are no different from those common in the housing, agricultural and commercial sectors.

Sometimes ponds are lined with plastics or other impermeable materials, and here the techniques are similar to those for civil structures such as potable water reservoirs or sludge tanks. The design and installation of water control gates, including in unstable soils, benefits from the long experience in this field in the agriculture and irrigation sectors. An understanding of hydrodynamics allows ponds and tanks to be built with good water circulation, oxygen mixing and without 'dead spots' where sediments might accumulate and cause health problems to fish.

For installations in the sea, the situation is somewhat different and many of the important engineering solutions, such as for fish cages or suspended shellfish growout systems, have had to be developed by aquaculturists themselves. They have benefited however from the accumulated knowledge of seafarers in general and fishermen in particular, in the design and operation of mooring and buoyage systems. More recently, when fish farmers have turned their attention to how to operate fish cages in locations further offshore where seas are rougher, the experience of the oil exploration industry has proven very valuable.

Modern Techniques

Techniques have been developed in recent years for the production of fish and other aquatic products in closed recirculation systems. To make these work, aquaculturists have needed to develop a knowledge of the biological processes operating — such as how bacteria can be used to neutralise and recycle the nitrogenous waste products produced by growing fish — and how to engineer the systems to meet the biological requirements. Knowledge of bioengineering from the waste treatment and water treatment industries has made contributions to the development of closed aquaculture systems and there is probably more that could be usefully transferred from the sewage treatment sector to help solve problems in fish rearing.

Modern materials have brought many benefits to aquaculture. For instance, custom made plastic joints have simplified the construction of sea cages, and made them more reliable in high stress conditions. Experiments with huge free-floating or sunken net cages operated in the open ocean were begun several decades ago, for instance in the Caspian Sea. These showed some promise, but more reliable construction materials will make the farming of fish in such structures increasingly feasible. Modern materials and production methods have been important also, in the construction of plastic filter substrates for indoor recirculating systems. The fine detail of these has been found to make substantial differences to the efficiency of biological filters.

In shrimp farms, specially designed matting materials that stand upright on pond bottoms, with a structure that promotes the growth of the small animals and plants that the shrimp can thrive on, have recently been developed and shown to boost production.

Engineering Skills

Engineering skills are important in the design of most aquaculture facilities and good engineering can affect the efficiency and economics of production. If capital costs can be minimised while still maximising productivity and reducing risk, the farming operation will be more profitable.

Aquaculturists have proven very innovative over the past fifty years, constantly developing new technologies to support their farming operations. As new production methods and species for farming are developed, the engineering solutions needed to support them will continue to evolve.

Quality Excavation and Construction of All Types

> *"The proposition of redoing an improperly constructed farm pond or fishing lake is always more expensive than the initial pond construction. Experience and knowledge count in this business."*

- Fishing Lakes
- Recreational Lakes
- Ranch Improvements
- Professional Project Consultation

- Farm Ponds
- Land Clearing
- Specialised Excavation Projects
- Specialised Lake Liner Projects.

All Fishing Lakes and Farm Ponds are NOT Equal!

A body of water on your property is a desirable feature which adds value and provides pleasure for the owner. Fishing Lakes are called many names.

In Texas and the Southwest, they may be known as pools, ranch lakes or stock tanks. In other areas of the country, it may be farm ponds, farm lakes, fishing lakes or swimming ponds.

Regardless of what the water feature on your farm or ranch is called, their use as water sources for livestock has diminished over the years. Today's property owner wants a farm pond, fishing lake or recreational lake which adds beauty, value and pleasure to their property.

Land, mother-nature's canvas. Completing the masterpiece requires an artist in the field of excavation. You need a contractor with imagination, a feeling for effect and form.

Figure: *Installing plastic lake liner near College Station, TX*

A client says "It is rare to find a consultant who puts such thought and effort into his work as Nick Jones. I recently had the privilege to witness his team of professionals build a beautiful lake, and sculpt my raw land into a much more aesthetically pleasing and functionally utilizable piece of property. Nick and his crew provide the quality craftsmanship that it takes to create a body of water that will enhance the value and enjoyment of your land."

Figure: *Driveway construction near Bryan, TX*

It takes a rare combination of experience, knowledge and proper equipment to complete a job at the highest possible levels. Most contractors never attain that level. Soilmovers LLC routinely exceeds our client's expectations.

5

The Need of Aquatic Plants

Aquatic plants are an often misunderstood and under-valued part of freshwater ecosystems. In fact, many people would rather not have them in their favourite swimming or fishing hole. The following points illustrate a few of the important roles that aquatic plants play in a water body.

Food

Aquatic plants provide important food for many animals. Ducks and geese eat the seeds, leafy parts, and tubers of plants such as pondweeds (*Potomogeten* spp.), watershield (*Brasenia schreberi*), arrowhead (*Sagittaria latifolia)*, water pepper (*Polygonum* sp.), and duckweed (*Lemna* sp.). Songbirds use fluff from cattails (*Typha* sp.) as nest material and eat the seeds of many emergent plants. Otter, beaver, muskrats, turtles, and moose will also graze on a variety of aquatic plants. More food for fish is produced in areas of aquatic vegetation than in areas where there are no plants. Insect larvae, snails, and freshwater shrimp thrive in plant beds. Sunfish—Minnesota's most sought-after game fish—eat aquatic plants in addition to aquatic insects and crustaceans.

Historically humans have also utilized aquatic plants as a food source. Cattails have edible shoots and roots and even the pollen has been used in making biscuits. Arrowheads form large edible tubers at the root ends, called duck potatoes, which were consumed by Native Americans. Watercress (*Rorippa nasturtium-aquaticum*) has many historic medicinal uses and its spicy vegetation continues to be used in salads and garnishes. Water lily roots are a common source of food in many parts of the world and have historic medicinal value. Even the submersed plant, coontail (*Ceratophylum demersum*) has been used for medicinal purposes.

Habitat

Aquatic plants provide important living space for small animals such as aquatic insects, snails, and crustaceans, which in turn supply food for fish and waterfowl. Many studies have shown that vegetated areas support many times more of these tiny creatures than do unvegetated areas. Plants provide shelter for young fish. Because bass, sunfish, and yellow perch usually nest in areas where vegetation is growing, certain areas of lakes are protected and posted by the DNR as fish spawning areas during spring and early summer. Northern pike use aquatic plants, too, by spawning in marshy and flooded areas in early spring.

Cover

Young fish and amphibians will use aquatic plants as a source of cover from predatory fish and birds. This, coupled with the abundant food supply, makes aquatic plants important nurseries for baby fish (including our native salmon), frogs, and salamanders.

Housing Supplies

The sturdy emergent plants provide nest and den-building materials for many birds and mammals, including muskrats. Humans also construct baskets, mats, boats, and even dwellings from cattail, rush, and bulrush stems.

Erosion Control

Submersed and emergent plants protect shorelines from erosion due to wave action or currents. They can also help stabilize the sediment which can increase water clarity.

Nutrient Cycling

Aquatic plants form a vital part of the complex system of chemical cycling in a waterbody. They can also influence the supply of oxygen in the water. Recently aquatic plants have received a lot of attention for their ability to soak up pollutants from contaminated water. They utilize nutrients that would otherwise be used by algae, thereby improving water clarity. Increasing attention is being paid toward their possible use as indicators of water quality.

Resist Invasion by Invasive Exotics

A diverse healthy native plant community is more resistant to invasion by opportunistic exotic plants.

Why are Aquatic Plants Seen as a Problem?

If aquatic plants are so wonderful, why are they perceived as a problem? Most of the time, problems arise when plants are so numerous they impede recreational activities such as boating and swimming. When growth becomes very thick, they also harm some fisheries, particularly juvenile salmon and trout habitat. The causes of unnaturally high levels of plant growth are complex. Often it is attributed to increased nutrients, which come from around the lake or in the watershed. Contributing problems can include failing septic systems, fertilizer run-off, or agricultural waste. These increased nutrients cause the natural process of lake aging (eutrophication) to proceed at an accelerated rate, and increased plant and algal growth is part of this process.

Another problem can arise if a nonnative species is inadvertently introduced to the lake. This often happens when recreational users unknowingly carry plants from one waterbody to another, or when someone discards aquarium plants into a lake. Several exotic species such as Eurasian watermilfoil (*Myriophyllum spicatum*) or Brazilian elodea (*Egeria densa*) are aggressive and can crowd out more desirable native vegetation. Changes in vegetation may take place slowly or quite rapidly.

Another problem can arise if a nonnative species is inadvertently introduced to the lake. This often happens when recreational users unknowingly carry plants from one waterbody to another, or when someone discards aquarium plants into a lake. Several exotic species such as Eurasian watermilfoil (*Myriophyllum spicatum*) or Brazilian elodea (*Egeria densa*) are aggressive and can crowd out more desirable native vegetation. Changes in vegetation may take place slowly or quite rapidly.

Solution

Along with preventing or eliminating pollution, you can monitor plant community changes by collecting and identifying aquatic plants on a year-to-year basis. This is also a good way to detect detrimental changes at an early stage when control or elimination of the problem is both less complicated and less costly. Collecting and preserving plants is not difficult, and the result is an increased awareness of aquatic plants, as well as a valuable historic record of what grows in the lake. However, proper identification of the plant can be tricky and is essential, particularly if you believe any to be exotic invasive species.

Benefits and Disadvantages of Aquatic Plants in Ponds

Aquatic plants are not considered desirable by many pond owners as they are viewed as a nuisance to such pond activities as swimming and fishing. However, all aquatic plants should not be viewed as undesirable, given the benefits they can provide. Many plants are critical to fish and wildlife communities associated with ponds because they provide cover, nesting areas, and food.

Ultimately, whether a plant provides a benefit or is considered a nuisance rests entirely with the pond owner and his/her goals for the pond. For example, a pond owner who desires a high-quality swimming pond often views any aquatic plant as undesirable and may go to great expense to eliminate these plants. Conversely, a pond successfully managed for wildlife will have a variety of aquatic plants present with only a few plant species being considered a nuisance.

The purpose of this fact sheet is to provide pond owners with insight into the value or negative aspects associated with various aquatic plant species.

If control of aquatic vegetation is the desired outcome, Ohio State University Extension Fact Sheets A-3-98, Controlling Filamentous Algae in Ponds, and A-4-98, Chemical Control of Aquatic Plants, should be obtained and read carefully. The University of Floridas aquatic plant web site at aquat1.ifas.ufl.edu/photos.html provides excellent pictures of most aquatic plants found in Ohio ponds.

Planktonic Algae

These algae are microscopic in size. Planktonic algae assemblages are commonly composed of diatoms, blue-green algae, and green algae, although other species can be found in Ohio. A better understanding of planktonic algae is available in Ohio State University Fact Sheet A-9-01, Planktonic Algae in Ponds.

Benefits

Planktonic algae are the foundation of the aquatic food chain in all ponds and lakes, and their abundance ultimately determines how many pounds of fish, such as largemouth bass and bluegill, can be grown in a pond.

Algal abundance is determined by water fertility with higher levels of phosphorus and nitrogen resulting in increased abundance. Higher planktonic algae density enhances food production for fish, making the pond capable of producing more pounds of fish.

As planktonic algae abundance increases, light penetration to deeper water becomes more limited. This, in turn, limits how far from shore submerged aquatic plants and filamentous algae can grow and whether or not they can or will become a nuisance for the owner.

As a rule of thumb, water clarity of 24 to 30 inches is desirable for fish production and for limiting other plant species abundances to tolerable levels.

Disadvantages

Too much of a good thing certainly can apply to planktonic algae if nutrient levels are too high. These algae can explode in response to high levels of nutrients and literally can turn a pond pea green, resulting in very low water clarity. In rare instances, pre-dawn oxygen levels can be reduced to lethal levels for aquatic animal life.

The ratio of nitrogen to phosphorus determines whether the planktonic community will be dominated by green algae that are readily eaten by microscopic animals or by blue-green algae that offer less value to the food chain. Low ratios of nitrogen to phosphorus favour blue-green algae rather than the more edible forms of green algae.

Filamentous Algae

Filamentous algae are microscopic algae that form colonies of filaments hence the name. These algae are notorious for forming the large, pillow-like mats of algae that float on the surface of ponds. Common types found in Ohio include Spirogyra and Pithophora.

Benefits: None in recreational ponds.

Disadvantages

As in the case with planktonic algae, high levels of nutrients can cause filamentous algae abundance to explode, especially in ponds lacking other aquatic plants, becoming so abundant that severe oxygen problems can result in the pre-dawn hours during July and August. Treating a severe filamentous algae problem in summer will almost certainly cause a fish kill. Ohio State University Extension Fact Sheet A-8-01, Winter and Summer Fish Kills in Ponds, provides insight into how these types of summer kills occur.

Excessive growth of filamentous algae ruins swimming in many ponds every year. No one enjoys swimming in a pond in which long filaments of algae cling to everything they touch. Additionally, anglers become frustrated with having to remove filamentous algae from their lures after every cast.

Submerged Plants

Submerged aquatic plants are common to most Ohio ponds, unless treatment with a herbicide has occurred or grass carp (white amur) have been stocked. Aquatic plants resemble terrestrial plants in many regards, but lack stem rigidity when removed from the water.

Benefits

Submerged plants are critical to a well-structured fish assemblage. They not only provide protection for small fish from predators but also produce large numbers of invertebrates for small fish, such as bluegill, to eat. Research has shown that the optimal abundance of submerged plants for largemouth bass-bluegill populations is 15 to 20% of the pondýÿs surface area. This proportion balances the predator-prey relationship between largemouth bass and bluegill so that both species have all sizes of fish represented in the assemblage.

Submerged plants also are an important food source for many species of waterfowl in the form of vegetation-dwelling invertebrates or the plants themselves. Notable duck species associated with submerged plant beds are blue-winged and green-winged teal, wood duck, gadwalls, American widgeon, and northern shoveler as well as several species of grebes. Many species of herons and egrets hunt the shallow areas of ponds for small fish and frogs where submerged plants occur. A pond managed for wildlife needs submerged plants as a habitat component.

Submerged plants also have an effect on water quality. Their ability to put oxygen into the water is an obvious contribution, but they also provide for long-term storage of nutrients that might otherwise be used to create nuisance levels of planktonic or filamentous algae. Ponds with beds of submerged plants have fewer problems with algae.

Disadvantages

As with all aquatic plants, an excessive amount of submerged plants can cause problems for the pond owner. The exception might be for the dedicated wildlife pond in which no other use is desired. When submerged plants are too abundant, they can cause problems similar to those described for filamentous algae.

Excessive amounts of submerged plants can present a problem for the pond owner who values his/her fishery. Once plant levels exceed 20% of the surface area, largemouth bass predation on bluegill becomes less effective. Small bluegill can effectively avoid being eaten by darting into the excessive vegetation. Thus, too many bluegill survive, and their growth declines due to increased competition from overcrowding. Bass growth also decreases, as they are unable to find prey effectively and consume enough to grow well.

The classic symptoms of this scenario are populations of small, thin largemouth bass and bluegill. Excessive submerged plants also pose a problem in ponds where water is being pumped out for irrigation, livestock watering, or might be pumped out in case of a fire. These plants can clog a pump intake, lessening the amount of water being pumped and shortening the lifespan of the pump due to excessive wear and tear.

Floating Plants: Floating aquatic plants come in all shapes and sizes from the pinhead size of a watermeal plant to the large leaves of a water lily.

Benefits

In general, watermeal and duckweed provide few benefits to a pond unless the pond is strictly a wildlife pond. The same species of waterfowl that eat submerged plants will also eat these small floating plants. Pond owners who value the natural, aesthetic qualities of a pond desire water lilies and their showy flowers. Largemouth bass and bluegill both like the shade provided by the large leaves of water lilies. The key is not to allow the lilies to overrun the ponds shallow areas, maintaining about 15 to 20% coverage

Disadvantages

As with other aquatic plants, an excess of nutrients can cause an overabundance of watermeal and duckweed. It is not uncommon for these small plants to completely cover a pondÿÿs surface in a few short weeks. This type of cover will substantially reduce light penetration so that oxygen-producing photosynthesis in the water underneath the floating plants ceases.

Warm water temperatures during summer months keep oxygen-consuming animal respiration and decomposition rates high, possibly resulting in a fish kill from lack of oxygen.

Water lilies are notorious for overspreading a pond in short order, especially if the pond is shallow and the water is clear. Clear water allows these lilies to send up stems and leaves from a greater depth.

This severely inhibits recreational activities. Many pond owners have planted water lilies for aesthetics, only to be frustrated with their quick coverage of the pond.

Emergent Plants

Emergent aquatic plants are plants that live in shallow water or just on the shore along a pond. Most of the vegetative portion of these plants grows above the water rather than under the water. From a distance, many resemble grasses. Table 2 provides a list of the emergent aquatic plant species common to Ohio ponds.

Benefits

Emergent plants provide important wildlife benefits to ponds. The plants themselves provide nesting habitat for a variety of bird species, such as song sparrow, red-winged blackbird, and wrens. Many other aquatic bird species, such as rails and herons, utilize emergent plant habitats during migration. Many mammals will make a meal of the green vegetative material growing above the water. Additionally, the seed heads of nutsedges, spikerushes, and bulrushes are attractive to many waterfowl species (mentioned previously) as a food source. A pond managed for wildlife should strive to have a variety of emergent plants.

Disadvantages

An overabundance of emergent plants can create a problem for some pond owners. This is particularly true for cattails, which account for nearly all pond owner complaints about emergent plants. Cattails have the ability to completely surround a pond and extend several feet into the water if allowed to do so. Even for a pond managed for wildlife, this poses a problem. Habitat diversity is good for wildlife, and a monoculture of cattails does not provide diversity. Excessive emergent plants can create problems for anglers who will find it difficult to effectively fish from shoreline areas. Cattails are highly attractive to muskrats, a mammal that can damage a pond in some circumstances. Dams are vulnerable to their burrowing activities, and muskrat burrows can compromise their integrity. Muskrat use cattails for a variety of reasons, including food, den material, and as an escape from predators.

One emergent plant species deserves special mention. Purple loosestrife is an invader to Ohio ponds and wetlands and should be controlled, either by pulling the plant and roots or by spraying a herbicide. If left uncontrolled, purple loosestrife quickly spreads and crowds out desirable native emergent plants. A monoculture stand of loosestrife is the result, providing no benefits to the pond owner or wildlife.

Explosion of Aquatic Plants

Massive occurrence of floating aquatic plants is a common feature on tropical and subtropical water bodies, and it is commonly the result of an expansive growth of non-endemic aquatic macrophytes. In a 1989 report to FAO, Hartley (CSIRO, Brisbane, Australia) estimated that two million people in Nigeria, including more than 24 000 fishermen, and approximately 100 000 persons living in riverine communities in Benin, who rely solely on fishing for their livelihood, are negatively affected by floating aquatic weeds (water hyacinth, salvinia, and water lettuce - *Pistia stratiotes*). Fishermen in Malawi and Ghana may be hampered by water lettuce. In the Niger, the more recent explosive growth of water hyacinth has interfered with fishing on floodplains of the Niger River. In lakes Kyoga and Victoria in East Africa, until very recently, thousands of fishermen were trying to keep open access to landing and fishing sites blocked by water hyacinth. Water hyacinth and salvinia are also widespread in the tropical zone of Asia and Latin America, representing a major problem for fisheries, irrigation, hydropower production and water transport.

Both the floating and submersed aquatic macrophytes can become nuisance (noxious) plants, especially where they appear as exotics. The problematic submersed plants are: *Myriophyllum spicatum, Hydrilla verticillata, Ceratophyllum demersum, Potamogeton crispus, Potamogeton pectinatus*. The reason for their explosive growth is often the favourable nutrient environment, and also human impact through eutrophication of some, especially small water bodies. The rapid expansion of the exotic water hyacinth in Sudan reduced the number of fish and fishing intensity (Davies, 1958/59). The periodical annual invasions of the Kerala (India) backwaters with salvinia renders fishing and boating impossible during certain months (Thomas, 1979). In the Sepik River in Papua New Guinea, where in the late 1970s and early 1980s salvinia spread over numerous floodplain lakes, eventually covering over 200 km, the tilapia (*Oreochromis mossambicus*) fishery collapsed (Mitchell *et al.*, 1980). The disruption of traditional subsistence lifestyle became so great that entire villages were abandoned (Anon., 1985). Fortunatelly, biological control using the beetle *Cyrtobagous salviniae*, which was released over the infested areas, was so effective that within only a few years salvinia virtually disappeared from all backwaters. Not always the presence of invasive plants becomes a nuisance. On Rawa Pening reservoir in Java, Indonesia, where all fish species are captured for food, control of water hyacinth led to overfishing, and only expansion of aquatic weeds due to the neglect of control measures led to rehabilitation of fish stocks (Carlander, 1980). Later on, a certain balance has been achieved among the water hyacinth area cover, capture fishery, and fish cage fish production.

The impact of invasive plants on fish stocks and fisheries is poorly documented, especially for the tropical and subtropical countries. Long-term monitoring is mostly lacking. Records available for the freshwater Lake Naivasha in Kenya, are an example of well-documented changes in a lake ecosystem under the impact of changes in aquatic macrophytes, but also of the impact of human interference with this freshwater lake.

Over the last 30 years, Lake Naivasha, situated at an altitude of 1 890 m, has experienced considerable changes in its plant as well as animal communities. *Salvinia molesta* appeared in the early 1960s, and toward the end of the 1960s it covered 2–3 km. Its appearance, combined with heavy fishing pressure on tilapia stocks, resulted in a decline to a level that the local frozen fillet plant had to be closed (Gaudet, 1976). Dissolved oxygen concentration underneath the mats was only 10 % saturation as compared to 64–85 % in the open water (Kongere, 1979). On Lake Kariba, Donnelly (1969) observed that tilapias established nurseries in shallow water only if it was free of *Salvinia*. Where salvinia mats shaded the water, phytoplankton production declined and the young tilapias moved out of the shallows into the deepwater *Potamogeton* and *Ceratophyllum* beds, where they were much more susceptible to predation. In Lake Naivasha the large mat of *Salvinia* almost disappeared by 1987, when it became limited to the river inflow along the northern shore (Harper *et al.*, 1990) Since 1991 the introduction of the biological control agent, the beetle *Cyrtobagous salviniae*, has kept salvinia under control (Abiya, 1996). The Lake Naivasha level fluctuations, with a high level lasting from 1960 to 1976, followed by a drop, which was followed in 1982 by a sharp rise in water level, has had considerable impact on other aquatic plants. In 1983 the water level started to decline again and by 1985 Kenya experienced a severe drought. By 1987 the lake level returned to the state of the 1950s. Papyrus (*Cyperus papyrus*) plants, formerly floating on the surface, became restricted to margins, but water-lilies (*Nymphaea caerulea*) started germinating, and extensive beds of submersed plants such as *Ceratophyllum demersum* and *Potamogeton* spp reappeared. The impact of these changes on fish and crayfish revealed that *Tilapia zillii* and *Oreochromis leucostictus* (both introduced species) showed resistance to the changes in lake level fluctuations and subsequent changes in aquatic plants. Both fish feed largely on detritus, which has always been in abundant supply.

In *T. zillii* macrophytes constituted less than 10% of the total food taken, while there were none eaten by *O. leucostictus*. Decline in tilapia fishery resulted from intensification of the fishery and the use of small-meshed nets. But whenever the lake water level rose, the flooding of terrestrial vegetation and shallows provided good nursery areas and this led to the recovery of fish stocks (Abiya, 1996).

Crayfish *Procambarus clarkii*, introduced in the lake in 1970 and another important component of the lake fishery, is a major consumer of aquatic plants. In the early 1980s it reached a density of about 4 crayfish m^{-2} in the littoral (Muchiri *et al.*, 1995).

The species is known to reduce aquatic vegetation in lakes in the USA at densities over 3 m^{-2} (Magnuson *et al.*, 1975). The crayfish fishery in Lake Naivasha crashed in 1983, this being followed by the recovery of rooted aquatics. This provides a circumstantial evidence that the decline and subsequent recovery of the native rooted submersed and floating aquatic plants was due to the heavy crayfish herbivory until 1983 and its relaxation thereafter (Harper, 1992). By 1987 the crayfish density was only 0.25 ind. m^{-2}.

Eichhornia Crassipes - Water Hyacinth

Water hyacinth is now widely spread throughout tropical and subtropical countries. Gopal and Sharma (1981) reviewed the available information on this plant for the years preceding their publication. They also provided maps on the global water hyacinth distribution to that date.

Rzoska (1976) described the invasion of *Eichhornia crassipes* in the Sudanese White Nile. The upper Nile swamps (the Sudd) have become a 'perennial' centre of infestation for the downstream Nile. In many water bodies of the Nile floodplains water hyacinth has fringed the shores in a belt of variable width, and it has taken over the function of *Pistia*. In fringing plants, up to a width of about 6 m, Bailey and Litterick (1993) found a rich community of aquatic macroinvertebrates inhabiting the root of this floating plant, and Hickley and Bailey (1986, 1987) found rich fish communities in water hyacinth fringes. Similarly, Green *et al.* (1976) found also rich fish fauna in water hyacinth fringes of Lake Lamorgan (Java, Indonesia). However, Davies (1958/59) and Rzoska (1976) noted that fisheries were impeded, specifically in areas inhabited by Shilluks, immediately north of the Sudd in Sudan, and on the Sobat River, also in Sudan. Gopal and Sharma (1981) estimated that in the 1950s about 45 000 t of fish were lost in West Bengal (India) due to the infestation of waters with water hyacinth. Referring to literature, they reported that fish production declined from 905 kg ha^{-1} at zero percent cover to 281 kg ha^{-1} at 25% cover, possibly due to low densities of phytoplankton as a result of shading by water hyacinth, and also the uptake of phosphorus by this plant.

One observation reported suppression of the phytoplankton production underneath 10% and 15% cover to 2.25 g m^{-2} and 1.97 g m^{-2}, respectively, as compared to 2.58 g m^{-2} at 5% cover (Gopal and Sharma, 1981). In experimental ponds in Alabama the production of the phytoplankton-feeding *Oreochromis aureus* was reduced by about 50% at 10–25% water cover by water hyacinth (McVea and Boyd, 1975).

In other situations, thick mats of water hyacinth cause complete depletion of dissolved oxyxgen content and consequently fish kills. The presence of water hyacinth in shallows makes such areas unsuitable for fish spawning. In the Sudd of Sudan, Bishai (1960/61) measured 1.8 mg L^{-1} of dissolved oxygen 30 cm below the mat when there was no water current, and he also reported high CO_2 concentrations, which increase the susceptibility of fish to low dissolved oxygen concentrations. Poi de Neiff and Solis (1994), who measured dissolved oxygen concentrations among the roots of water hyacinth floating islands on the Parana River floodplain in Argentina, recorded a maximum of 2.3 mg L^{-1} within the first 1 m depth, but the usual concentrations were close to 1 mg L^{-1}, declining with depth.

Two Amazon River floodplain fish species, *Colossoma macropomum* and *Schizodon fasciatum*, have shown considerable resilience to hypoxic conditions. Experimental investigations demonstrated that if the hypoxic conditions persist, these species return to the zones of plant cover and remain among the roots of floating plants, where the oxygen availability among the roots seems to be better than in the deeper water. In experiments with *Eichhornia crassipes* and *Pistia stratiotes*, conducted by Jedicke *et al.* (1989), the roots of these plants released oxygen into the water. The input of oxygen beneath a layer of *E. crassipes* was nearly double that beneath an open water surface. The input from water lettuce (*P. stratiotes*), however, was only slightly greater than that determined in open water. Jedicke *et al.* (1989) has pointed out the advantages for the hypoxia-tolerant fish species to remain among the plant roots: reduced predatory pressure; utilization of food supply when other species cannot access it; lower danger to fish from birds and other terrestrial predators, to which fish species displaying emergency respiratory behaviour at the surface of open water zone are exposed. Bailey and Litterick (1993) found markedly less aquatic macroinvertebrates inhabiting the roots of water hyacinth when appearing as small floating mats, as compared with those of the fixed fringe, and attributed this reduction partly to the feeding activity of fish. It is probable that if the plant appears in broken small-sized mats, which allow to maintain dissolved oxygen concentration underneath the mat at reasonably high levels, its is sought by fish to feed on its periphyton and the associated aquatic invertebrates. The plant also provides a habitat to invertebrate hosts and vectors of parasitic diseases such as malaria, schistosomiasis, encephalitis and elephantiasis.

In the 1970s and 1980s in East Africa water hyacinth invaded the lakes Kyoga, Kwania and Victoria. In Lake Victoria water hyacinth was noticed for the first time in 1988, and by 1992 it was already well established (Ogutu-Ohwayo, 1995). In the 1990s it has been recorded also from Lake Albert. It has had a major negative impact on fish catches and fish landings, due to its obstruction of fishing and landing sites. It is also believed that its presence along the shores has had a major impact on tilapia nesting, through lowering the dissolved oxygen above such sites and shading. In the early 1990s the worst affected areas have been on Lake Kyoga (FAO, 1993a), but by 1996, in the Uganda sector of Lake Victoria, the weed extended an average of 15 m from the shore, covering almost 80% of the country's 1 000 km of shoreline. Near Kampala, the capital of Uganda, the weed was 30 m wide. Prevailing winds from the south pushed most of the weeds to Uganda, but in 1996 the weed almost encircled the 3 000 km of shoreline (Anderson, 1996). Luckily, the situation has dramatically changed for the better, and by January 2000 it was reported that water hyacinth had been reduced by 80% on Lake Kyoga and 70% on Lake Victoria (Wendo and Ngatya, 2000). In Lake Victoria only 16% of the shoreline remains infested. This reduction is believed to be the result of the success of biological control using two weevil species, i.e. *Neochetina bruchi* and *N. eichhorniae*, released on Lake Victoria since 1996, and on Lake Kyoga since 1994.

With the decimation of haplochromine stocks by Nile perch (*Lates niloticus*), especially in Lake Victoria, a potential benefit resulting from the presence of water hyacinth has been observed: this floating plant appears to provide cover for haplochromine stocks which have been on the increase since the appearance of water hyacinth in Lake Kyoga. In the reservoir Rawa Pening in Java, Goeltenboth and Kristyanto (1987) observed that water hyacinth is a valuable habitat for fish as long as it does not form extensive solid mats.

Ogutu-Ohwanyo (1995) in his comments on the potential for controlling water hyacinth in Lake Kyoga implies that reduction in the water hyacinth cover might again expose haplochromines to Nile perch predation. The haplochromine-aquatic macrophyte relationships are far from clear. Witte *et al.* (1992) suggested that continuous digging by haplochromines in Butimba Bay, Lake Victoria, could have prevented the settling of rooted plants such as *Nymphaea* spp. With the decline in haplochromines, since the end of the 1980s, the shallows in this bay have been overgrown by *Nymphaea* spp, *Ceratophyllum demersum* and invaded by water hyacinth. Witte *et al.* (1995) believed that water hyacinth invasion of Lake Victoria threatened haplochromines and other fish. It would be interesting to know the result of the presently rapidly diminishing water hyacinth cover on haplochromines and other littoral fish.

In Ghana, water hyacinth was found for the first time in Tano lagoon, apparently entering this water body from Cote d'Ivoire. More recently fishermen reported a decline in fish catches and it was suggested that this was due to the presence of the water hyacinth. Much of the water hyacinth in Tano lagoon is mobile, being moved by wind or water currents. Holcik (1995), who investigated the situation, thought that the low catches are not due to the decline in fish stocks but rather a result of both the inefficient deployment of nets and of the difficult access to the fishing grounds due to the water hyacinth cover.

In Laguna de Bay, a large lake near Manila, the capital of the Philippines, infestation by water hyacinth has been a major environmental problem in several areas of this lake. Massive proliferation of the plant not only poses a hazard to navigation, but it also leads to poor water quality in this shallow lake, which has considerable fisheries importance. Tilapia and milkfish have been grown in numerous fish pens, the area of which at its maximum expansion reached almost 30 000 ha, representing about one third of the total water surface. Frequent typhoons have been known to drive the floating plants against fish pens (Photo 6), destroying them and releasing the penned fish into the lake. In the late 1980s water hyacinth cover was estimated to be 1 800 ha, and the plant was also present in tributary rivers, creeks and irrigation canals. This resulted in difficult, if not impossible access to fish pens, fishing sites, damage to fish pens, fishing nets and other fishing gear, reduction in fish population due to the low dissolved oxygen levels and increased carbon dioxide concentrations, blockage of water intakes of irrigation pumps, and interference with water transport. The removal of water hyacinth was largely manual, plus using one plant harvester. Monitoring of water hyacinth infestation in Laguna de Bay was done with the help of remote sensing using Landsat imageries (Bina *et al.*, 1978).

Tin-mine lakes, of which about 4 300 exist in Malaysia, are a consequence of abandoned tinmining pits, which have filled with water. Some of them have been used for aquaculture, and these represent 18.9% of the total area used for freshwater fish pond culture in peninsular Malaysia (Arumugam, 1994). While a total of 55 species of aquatic macrophytes have been identified from these lakes, in such lakes which receive a significant nutrient input, drastic changes in macrophyte and other communities take place. Once fish culture has been abandoned, in those lakes which are used as dumping sites for organic waste, *Eichhornia crassipes* may cover up to 90% of the surface area. Such lakes contain mainly air-breathing fish (*Channa* spp, *Trichogaster pectoralis*) which tolerate low levels of dissolved oxygen underneath the mat of floating vegetation. Lakes lacking the dense cover of floating macrophytes support indigenous fish and tilapias.

The widespread abundance of water hyacinth has attracted the attention of fish feed specialists and fishery managers. Water hyacinth has been tested both as a direct food of fish, and as a fish feed component. For example, in experiments carried out in Java, Indonesia, up to 10% of the feed was either composted or non-composted water hyacinth. It was noted that the feed, fed to *Puntius javanicus* and *Cyprinus carpio*, did not slow down the growth rate of these fish. *Trichogaster pectoralis* has shown a significantly slower growth rate when fed higher levels of substitution (Hutabarat *et al.*, 1986). Numerous other experiments have been performed on the use of water hyacinth as an additive to fish feed, but they are not a subject of this review.

In Indonesia, water hyacinth, together with some other floating plants, is used as a fish attracting device. In shallow lakes, an enclosure made of bamboo poles pushed into the bottom mud, and filled with aquatic plants, is called *'bungka todo'*. Tree branches or brush are often added to the aquatic plants. In Kalimantan, such systems are used on inland lakes of the Mahakam River in the east. On the Kapuas

River in the west, water hyacinth is rarely found as it does not grow well in black waters of inland floodplain lakes, therefore any water hyacinth which is found is treasured and used as shade to protect fish in cages from direct sunshine, rather than using it as *bungka todo* (Petr, 1993) (Photo 8). The plants provide the fish with shade and food, especially periphyton and its associated invertebrate fauna. The floating aquatic plant enclosures are regularly fished. It has been suggested that in lakes threatened by overfishing, such as Lake Tempe in Sulawesi, *bungka todo* could be an efficient way of fish stock protection and preservation in areas set aside and designed strictly off-limits for fishery. From them, the unprotected areas of the lake would then be naturally recolonized, at least by some fish species.

Pistia Stratiotes

Pistia stratiotes is a free floating common aquatic plant in tropical Africa, where it occasionally forms dense and large mats. In most situations, such mats are not permanent, and are moved or dispersed by wind. In some water bodies, *Pistia* may become part of sudd. The plant consists of a rosette of leaves covered with hydrofobic hairs extending above the water surface, and of submersed, finely divided roots reaching to a depth of 15–20 cm. The roots usually harbour a rich community of invertebrate fauna, which in turn is preyed upon by fish and aquatic predatory invertebrates, such as dragonfly nymphs.

In Volta Lake (a reservoir in Ghana, West Africa), large floating mats of *Pistia stratiotes* developed soon after the reservoir started filling. It became abundant especially where shelter against wind and waves was provided by floating or stranded trunks of dead trees, and in between dead trees standing in shallows. In places it was knitted together with *Scirpus cubensis*.

The biggest expansion of *Pistia* carpets occurred toward the start of rising water level. After the water level stabilized, *Pistia* became less numerous, with large plants gone, and only small plants remaining. In *Pistia* plants, situated well inside a mat, roots contained between 5 000 and 16 000 m^{-2} aquatic macroinvertebrates (5–35 g m^{-2} wet weight), excluding zooplankton (Petr, 1968).

The values are lower than those for the margins of floating meadows of varzea, where Junk (1973) recorded a maximum of 780 000 individuals (62 g m^{-2}). Junk included the very abundant crustaceans (Copepoda and Cladocera) in his samples. Throughout the observation period, in Lake Volta the *Pistia* roots macroinvertebrate fauna was dominated by the predatory dragonfly nymphs, which although low in numbers exceeded browsers and filter feeders in biomass. The presence of macroinvertebrates was considered to be of importance for fish feeding on periphyton and on it associated invertebrates. In 1965, the first year of the reservoir filling, *Pistia* of Lake Volta also harboured the snail *Bulinus forskalii*, which was very numerous at the peak of the dry season, when the water level was slowly declining and the surface water temperatures were at their highest. In 1966 the first *Bulinus globosus rohlfsi*, a host of schistosome (*Schistosoma haematobium*) parasites, was recorded from the lake (Paperna, 1970) while *B. forskalii* became rare. In one area Paperna collected over 50 *B. t. rohlfsi* from a single *Pistia* plant. When in some areas of the lake *Pistia* started disappearing, the snails thrived particularly on *Ceratophyllum demersum*, which became very common. This change has led to a dramatic increase in the incidence of schistosomiasis among the fisherfolk on this reservoir.

Little is know about the fish feeding on the aquatic fauna and periphyton associated with the submersed parts of *Pistia*. In Lake George, Uganda, experimental catches within and alongside drifting mats of *Pistia*, which entered a bay surrounded by papyrus, haplochromids represented 98.3% of the total number, and 98.2% of the total biomass of fish captured during a 24 hour period (Petr, unpublished). Among the haplochromids *Haplochromis nigripinnis*, *H. aeneocolor*, and *H. angustifrons* represented 66.5, 17.8, and 13.2% by number of the total, respectively, and a biomass in a similar proportion. *H. nigripinnis* is an omnivore feeding predominantly on algae, insect larvae and emerging insects; *H. aeneocolor*, which is abundant along the papyrus edge, feeds on insect larvae and adult aquatic insects, and the guts contain also plant fragments, suggesting that this species may selectively forage on the food among the roots of papyrus and *Pistia* and that the plant fragments are roots taken incidentally with the insects; *H. angustifrons* is predominantly an open lake species feeding on insect larvae, especially chaoborids, and on zooplankton (Dunn, 1975).

Pistia may be used as a shading plant in aquaculture. In tropical Australia, it has been used in a raceway culture of juvenile growout of freshwater crayfish *Astacus quadricarinatus*. The use of this plant was considered advantageous for several reasons: provision of suitable habitat/substrate for juvenile crayfish; photosynthetic production of oxygen; stabilization of water temperature; absorption of nitrogenous wastes; provision of natural food organisms associated with the plant, i.e. epibionts; shading effect; high carrying capacity for crayfish juveniles with reduced intra-specific competition; enhancing survival and growth of the juveniles (Jones, 1988).

Salvinia Molesta

Exotic aquatic fern, the floating *Salvinia molesta* is spread widely through the tropics. Thick mats of this plant prevent or make difficult boat traffic (Photos 9 and 10), impede or make fishing impossible, and the plants harbour also invertebrate hosts of water-borne diseases.

Solid mats of salvinia depress dissolved oxygen concentration, increase concentration of carbon dioxide, and occasionally may result in formation of hydrogen sulfide in water underneath it. Lake Kariba, a large reservoir on the Zambezi River, was invaded by salvinia shortly after the formation of this reservoir.

After some years salvinia, covering at its largest expansion in 1962 approximately 1 000 km^2, i.e. 23% of the reservoir surface area (Karenge and Kolding, 1995), declined in the 1970s to about 1% (Mitchell and Rose, 1979), largely due to nutrient deficiencies.

Salvinia was replaced by increasing amounts of rooted macrophytes down to a depth of around 10 m. In parallel with that there was an increase in the benthic macroinvertebrates and in benthic-feeding fish species such as *Mormyrus longirostris*, *Synodontis zambezensis*, *Serranochromis codringtonii*, *Serranochromis macrocephalus* and *Labeo altivelis*. Increased diversity within the inshore fish community in Lake Kariba was attributed to the development of a mosaic of new niches among the aquatic macrophytes, periphyton and submersed trees where a variety of benthic gastropods, molluscs, insects and other invertebrates provided a new food resource to fish (Karenge and Kolding, 1995). The lake littoral is, however, not a stable environment due to a drawdown of this reservoir, which serves the production of hydroelectric power for Zambia and Zimbabwe. It was recommended that to promote full development of macrophytes, the drawdown should not exceed 3.5 m. This limit is sometimes exceeded in Lake Kariba, where dry years may irregularly alternate with wet years. In spite of that, the development of a macrophyte belt has enhanced the stocks of juveniles and small fish species such as *S. macrocephalus*, *Brycinus imberi* and *S. zambezensis* which have increased during the 1980s (Karenge and Kolding, 1995). The authors also mentioned that in parallel with the increase of the above fish species there was a decline in small-sized *Hydrocynus vittatus*, *Hippopothamyrus discorhynchus* and *Marcusenius macrolepidotus*. The reason for this decline is not understood.

With respect to aquatic macroinvertebrate populations on *Salvinia molesta* as compared to those on water hyacinth, Arunachalam *et al.* (1980) reported a higher diversity from the first plant, where they found 66 species, than on the roots of water hyacinth (about 30 species) from the freshwater lake Veli in India. The authors do not provide information on the biomass of the macroinvertebrates on these two plants, hence it is difficult to assess which plant is more important for fish as a source of invertebrate prey.

In Lake Kariba, salvinia rootlets were found in stomachs of a great variety of fish species of herbivorous, omnivorous and even piscivorous character. In *Tilapia rendalli* they were found in 19.6% of fish stomachs, but represented only 6.7% of the total food taken (Mitchell, 1976). Salvinia rootlets, detritus or fragments occurred also in: 9.3% *Haplochromis codringtoni*, 11.1% *Haplochromis darlingi*, 19.2% *Pseudocrenilabrus philander*, 29.6%, *Eutropius depressirostris*, 8.1% *Clarias gariepinus*, 24.0% *Synodontis zambezensis*, 38.2% *Micralestes acutidens* (where it formed 19.2% of the total content), 12.3% *Alestes lateralis*, 11.1% *Mormyrus longirostris*, 5.0% *Marcusenius macrolepidotus*, 10.2% *Hippopotamyrus discorhynchus*, 8.1% *Mormyrops deliciosus*. In the majority of fish species it would appear that salvinia rootlets and detritus were taken when feeding on aquatic macroinvertebrates and plankton residing among the rootlets. For example, out of the 19 fish species investigated for food, 15 contained the crustaceans *Cyclestheria* and *Caridina*, 18 contained the nymphs of the mayfly *Povilla adusta*, 14 contained chironomid larvae, 13 contained trichopteran larvae.

In Sri Lanka, salvinia is often removed from water reservoirs manually, but biocontrol using the beetle *Cyrtobagous salviniae* is more efficient.

Cyperus Papyrus

Papyrus rhizomes are typically rooted in loose detritus in a floating mat. Papyrus may occur as free-floating or stranded, or forming fringe swamp. Occasionally, the lake-side edge of the fringe papyrus swamp will float if the lake level permits.

The extent of papyrus swamplands in East and Central Africa is enormous. In southern Sudan the Upper Nile forms large swamps called the Sudd, which covers approximately 40 000 km^2 (Rzoska, 1974). In Uganda, papyrus encircles completely Lake Kyoga, and is well developed in Lake Victoria basin.

In Lake George, Uganda, where it forms large fringing floating swamps, papyrus flourishes in water with a conductivity of 200 ìS, while in Lake Edward (850 ìS) it is restricted to river mouths (of much lower conductivities) on the Uganda shore of the lake (Lock, 1974). Gaudet (1980), in summarizing the significance of the papyrus for the aquatic fringe ecosystem, pointed out that papyrus swamps supply large amounts of fixed nitrogen to tropical lakes and rivers every year thus allowing an increase in animal and plant production at the swamp edge. Papyrus fringes act to moderate changes along a river system by regulating nutrient flow and recycling. For the most part this is accomplished by production of an organic nutrient output at the expense of an inorganic nutrient input. Papyrus plants effectively extract dissolved nutrients from tropical river systems, but such nutrients are later exported as organic particulate matter or adsorbed to particles which are carried into a lake by through-flow (Gaudet, 1979). In practical terms, papyrus swamps may be effective in removal of organic nutrients from sewage effluents into these swamps (Gaudet, 1978), since papyrus seems to be quite tolerant to pollution.

The organic nutrients released from the organic particulate matter or adsorbed to particles represent an energy source to a very diverse tropical fauna and flora, which in turn supports a profitable shallow water fish industry in many parts of Africa (Gaudet, 1980). Summarizing the importance of papyrus swamps along the major rivers in Africa, such as Congo and White Nile, Gaudet characterized them as an ecological buffer zone, which moderates changes along the river system by regulating nutrient flow and recycling. For the most part this is accomplished by production of an organic nutrient input at the expense of an inorganic nutrient input. The organic nutrients serve as an energy source to a very diverse tropical fauna and flora, which in turn supports a profitable shallow water fish industry in many parts of Africa. At lake edges papyrus swamps are often sites of intensive fishing and thus a resource worthy conservation.

For air-breathing fish papyrus swamps and other large swampy divides are not likely to be a barrier to dispersal. For water breathers, the use of and dispersal through papyrus swamps is likely to be limited by the availability of oxygen, the open surface area, and efficiency of oxygen uptake in the species (Chapman and Liem, 1995). In a study of a papyrus swamp, which feeds a small river in Uganda, Chapman *et al.* (1998) found peak of dissolved oxygen concentrations associated with biannual flooding, and they suggested that this provides opportunities for fish to use or move through papyrus. However, peak flood values never reached 50% saturation, and thus may pose barriers for some hypoxia intolerant species. At the lake edge of Lake George in Uganda dissolved oxygen showed a rapid decrease moving inside the papyrus belt (Petr, 1973). Chapman *et al.* (1998) reported *Ctenopoma muriei, Clarias liocephalus* and *Nothobranchius* sp. to be found principally in papyrus swamps and other wetlands of the Lake Victoria Basin. Chapman and Liem (1995) found the small cyprinid *Barbus neumayeri* to inhabit the dense interior of papyrus swamps, due to the presence of large gills and the use of aquatic respiration at the air-water interface. Chapman *et al.* (1995) reported that 12 species of East African cichlids are tolerant of low dissolved oxygen concentration and able to use low oxygen habitat along the margins of Lake Victoria as a refugium against the introduced Nile perch.

In lakes with a more pronounced fall in water level, papyrus stands may become drastically reduced. In Lake Naivasha water level declined by just over 4 m between 1980 and 1987, exposing some 6 500 ha of soils, or about 35% of its 1980 area (Harper *et al.*, 1993). During this decline exposed soils were cleared of vegetation, mostly stranded *Cyperus papyrus* and other emergent plants, and put down to agriculture. Harper *et al.* (op. cit.) noted that fringing papyrus is now voluntarily kept by some landowners to keep a buffer zone, which assists in reducing the input of high nutrient loads into the lake. The increasing demand for water is seen as a direct threat to the lake in the future (Abiya, 1996). For details on water chemistry in papyrus swamps and underneath floating papyrus mats of the fringing zone, and Thompson (1976).

Submersed Aquatic Plants

Myriophyllum spicatum, a native Euroasian plant, spread rapidly across North America in the 20th century, displacing native species in some water bodies and often forming dense stands and surface mats. In Currituck Sound in North Carolina, USA, it was detected for the first time in 1965, when it covered 40 ha. In 1966 it covered 3 200 ha, and by 1974, 32 000 ha were infested (Sheldon, 1997). *M. spicatum* can withstand a broad range of conditions: it can grow in oligotrophic to eutrophic waters, in 0.5 m to 8 m depth, in sand, in organic substrate, under pH ranging from 5.4 to 10, in brackish water, in northern temperate lakes and rivers, and in subtropical water bodies.

In 1978 in Lake Pontchartrain estuary, this macrophyte has become a dominant species, with a lower abundance of the native macrophytes *Vallisneria americana* and *Ruppia maritima* present there. Duffy and Baltz (1998), who in the estuary studied the impact of the invasive *Myriophyllum* on fish, found that the highest fish species diversity was in *Vallisneria*, while the lowest in *Ruppia* and *Myriophyllum*.

The common fish were significantly more abundant in vegetated areas than on adjacent unvegetated substrate, and total abundances were higher in *Myriophyllum* and *Ruppia* than in *Vallisneria*.

The authors suggested that the exotic *Myriophyllum* may not have had a detectable influence on fish assemblages or abundances relative to the native macrophytes because the disturbance of the substrate by waves in more open areas prevent it from growing densely enough to strongly alter the original microhabitat. However, in some other water bodies, *M. spicatum* may form thick, impenetrable walls of uniform height and leaf form. When the stem density exceeds 250 stem m^{-2} fish start avoiding the plant (Savino and Stein, 1982). Keast (1984) found a lower fish density in *Myriophyllum* beds than in native aquatic macrophytes, and Dvorak and Best (1982) found lower densities of aquatic macroinvertebrates in this plant, when compared to native plants.

Sloey *et al.* (1997), who studied the distribution of aquatic macroinvertebrates in beds of *M. spicatum* in Fish Lake, Wisconsin, found higher invertebrate biomass, density, and taxa richness in the upper foliated sections of plants versus the lower unfoliated sections. The edges of the beds contained higher biomass, density, and taxa richness than the centre of the bed. They did not study the impact of the differential distribution of macroinvertebrates on fish, but other studies confirmed that the structural complexity of plant species and spatial distribution of plants are among the decisive factors determining which and how many fish are frequenting aquatic macrophyte stands.

Hydrilla verticillata is another invasive submersed species, which is present in numerous water bodies of the Americas, Europe, Africa, Asia and Australia. In some water bodies in the USA this plant has led to a considerable increase in fish stocks and fish species of sport fishery importance, while in other water bodies its presence has required a management response to prevent decline in recreational fishery. Apart from grass carp, for which this plants is a preferred food, other fish have also been used for controlling this plant. In Africa, *Tilapia zillii* and *T. rendalli* are known to feed readily on it, in India *Cirrhinus mrigala* and *Etroplus suratensis*, stocked in nursery tanks with *Channa punctatus*, efficiently controlled this plant (Raj, 1976), and also in India, *Tor tor* and *Puntius* spp feed readily on it. Herbivorous turtles in Bangladesh, and manatee in the USA and some Latin American countries also feed on hydrilla.

Menace of Excessive Aquatic Plants

Water hyacinth (*Eichhornia crassipes*) which have been described as the most troublesome weed in the world have been linked to several problems like obstruction to water transportation, micro-habitat for disease vectors, obstruction to fishing and reduction in biodiversity. However, recent studies have also shown that this macrophyte can be used for the production of paper, biogas, fertilizer, fish feed and in the clean-up of polluted environment (phytoremediation). It becomes important to fully harness the potentials of this aquatic macrophyte, which could change its status from a weed to an income-generating plant.

The aquatic macrophyte called water hyacinth (*Eichhornia crassipes* [Mart.] Solms) is not new in the ecological history of man. Infect, it has been popularly described as the most troublesome weed of the world because of its rate of multiplication. It is aquatic in nature, deriving its energy form the sun, storing this energy in a semi-succulent stem and a fibrous network of roots.

The commonest water hyacinth is the South American species-*Eichhornia crassipes*. It is distinctly divided into three (3) parts; a fleshy leaf (leaves) that form the basis of photosynthetic accumulation, a greenish semi-succulent stem and a rather brownish fibrous root network. The latter is usually submerged under the water and forms the propellant part of the weed.

As a crop, the weed is still a mystery to many ecologists by virtue of its rapid multiplication. To say that it breeds like a rat is an understatement, to compare it to the hydra or *Paramecium* could be an exaggeration. However, the weed has a high multiplication factor. *Eichhornia crassipes* has as much as seven-fold (7) increase or spread in 50 days time. Infact, it has been reported that 2 hyacinth plants can grow into 1,200 offspring in 120 days.

A possible biological explanation to this behaviour lies in the fact that the plant grows usually in places with high solar energy constants, ranging form 450 to 550 watts per metre square of land. Coupled with this is its high photosynthetic fixation efficiency of 1.52%. When compared to typical crops in sub-saharan Africa like maize, with a value of 1.0%, cocoa with value of 0.5%, elephant grass with a value of 0.48% and groundnut with value of 0.29%, one sees immediately that water hyacinth has a rapid growth tendency.

Cultivation reports from some swampy areas of South America have it that the plant has an annual average productivity of about 350-1,700 tons per hectare of wet vegetation. With this high growth rate, waterways are easily clogged up, because "clearing" rate is far below the "multiplication" rate of the plant.

This present "menace" position could turn to be a blessing in the near future as possible utilization are sought for the seaweed in the production of animal feeds, detergents and recently in phytoremediation; the deliberate introduction of aquatic weeds like water hyacinth into polluted water bodies so that the seaweed so introduced will absorb pollutants like nitrate, phosphate and heavy metals and thus, purifies the water.

Scientific Classification of Water Hyacinth

Kingdom	:	Plantae
Division	:	Magnoliphyta
Class	:	Liliopsida
Order	:	Liliales
Family	:	Pontederiaceae
Genus	:	*Eichhornia*

Species

E. crassipes	:	Common water hyacinth
E. azurea	:	Anchored water hyacinth
E. diversifolia	:	Variable leaf water hyacinth
E. paniculata	:	Brazilian water hyacinth

Elemental Composition of Water Hyacinth: Fresh water hyacinth contains about 90% water and about 15-20% solid materials. On a dry weight basis, the weed contains about 25-35% protein-related matter. Specifically, amino acids constitute about 17% of the total protein matter, the rest being amides. This explains why water hyacinth cannot be eaten fresh like most other edible green vegetables like lettuce (the amides are usually toxic matters).

Further analyses show that the carbon content of the dry weed is about 36-40%. Direct carbonylation reactions reveal that the carbonates and nitrates obtained were in the order of 40 to 60% in yield ratios. Thus, water hyacinth has a predominantly cellulosic structure highly impregnated by the amino group directly at the carbonyl structure. The aquatic plant can be conveniently represented as:

Investigations carried out have revealed R to be more of the aliphatic chain. Summarily, therefore, the water hyacinth in elemental composition is about 12.8% nitrogen content, 36-40% carbon, hydrogen about 8% and oxygen about 13-14%.

The aquatic plant is also known to be "carrier" of heavy metals like iron, magnesium and zinc, which suggest why it's being used in photoremediation. There are negligible traces of phosphorus and calcium. These metallic ligands can possibly account for the rest of the structure.

Description of Water Hyacinth: Water hyacinth *(Eichhornia crassipes*) is a member of the pickere/weed family-Pontederiaceae. The plants vary in size from a few centimeters to over a metre in height. The glossy green, leathery leaf blades are up to 20 cm long and 5-15 cm wide and are attached to petioles that are often spongy-inflated. Numerous dark, branched, fibrous roots dangle in the water from the underside of the plant.

The inflorescence is a loose terminal spike with showy light-blue to violet flowers (Flowers occasionally white). Each flower has 6 bluish-purple petals joined at the base to form a short tube. One petal bears a yellow spot. The fruit is a three-celled capsule containing many minute, ribbed seeds.

Geographic Distribution: Water hyacinth, "perhaps the world's most troublesome aquatic weed" is a native of tropical South America that has spread to more than 50 countries on five continents and has become a massive problem in waterways in both Africa and Southeast Asia. Its air-filled tissue (aerenchyma) enables it to float and spread rapidly within and between connected water bodies. It reproduces asexually by breaking apart into pieces each of which develops into a separate plant. This results in a rapid increase in biomass and continuous mats of living and decaying water hyacinth up to two metres thick covering the water surface have been reported.

Water hyacinth was introduced to North America in 1884 via the Cotton States Exposition in New Orleans. The plant was displayed in ornamental ponds and distributed as souvenirs to visitors, with the excess dumped into nearby creeks and lakes.

The plant soon overtook these lakes and creeks and controlling it became a problem. According to Joyce (1992), these problems led to passage of the River and Harbour act in 1899, authorizing the United States Army Corps of Engineers to maintain navigation channels in these areas. Control efforts included the spraying of sodium arsenite, which poisoned applicators and livestock. In Nigeria, water hyacinth was first noticed in 1985 in Lagos Lagoon. Since then, it has spread to a lot of water bodies in the country, making navigation and fishing an almost impossible task. Water hyacinth is found in a lot of water bodies in the tropics.

Habitat: Water hyacinths grow over a wide variety of wetland types form lakes, streams, ponds, waterways, ditches and backwater areas. Water hyacinths obtain their nutrients directly from the water and have been used in wastewater treatment facilities. They prefer and grow most prolifically in nutrient-enriched waters. New plant populations often form from rooted parent plants and wind movement and current help contribute to their wide distribution. Linked plants form dense rafts in the water and mud. In the Pacific Northwest, water hyacinth is planted outdoors in ponds and aquaria, but it is not considered winter hardy, except under special conditions.

Reproduction: Water hyacinth (*Eichhornia crassipes* [Mart.] Solms) reproduces sexually by seeds and vegetatively by budding and stolon production. Daughter plants sprout from the stolons and doubling times have been reported by Westerdahl and Getsinger (1988) as 6-18 days. The seeds can germinate in a few days or remain dormant for 15-20 years. They usually sink and remain dormant during periods of stress (droughts). Upon reflooding, the seeds often germinate and renew the growth cycle.

Control of Water Hyacinth: There are three common methods that have been used to control water hyacinth. These are:

- Mechanical control method
- Chemical control method
- Biological control method.

Mechanical Method: Ever since it colonised the Northern Rivers of New South Wales with such devastating speed, water hyacinth has tested human ingenuity in devising control techniques. Early control attempts concentrated on removing the plant from water with hand or instrument like pitchforks, then dumping the accumulated mass on land to die. This control method is costly in time, money and energy and several of the procedures used damaged the ecology, affecting all animal life in the aquatic ecosystem infested by the hyacinth.

Chemical Method: This method involved the use of herbicides to control water hyacinth. Various kinds of herbicides such as 2, 4-D, Dalapon, Diquat and others have been used in some places. They also opined that the ecological problems created by these herbicides were obvious.

The water could not be used for irrigation or human consumption for long period of time and the fauna in the ecosystem were seriously affected. Raju and Raddy working on some herbicides concluded that the best chemical control was achieved with 2, 4-D dimethyl amine 58% (4 kg ha^{-1}). They also reported that the cost of removal by this herbicide was 61% less than that of manual weeding.

Biological Method: Biological control of water hyacinth has been studied with several kinds of animals like viruses, bacteria and fungi as well as with Manatees, insects, herbivorous fish such as grass carp and Tilipia, ducks, geese; turtles, snails and other animals. However, the results were disappointing, perhaps because of defense mechanisms in the plants.

For example, the larger plants form 2.5 or more leaves for each one destroyed by pathogen attack. Two insects that have enjoyed the widest usage in the biological control of water hyacinth are-*Neochetina eichhorniae*-aweevil and *Neochetina bruchi*-a weevil also. A moth; *Sameodes albiguttalis* have also been used. Jayanth (1988) used *Neochetina eichhorniae* and got an impressive result where 95% of the infestation was cleared within 32 months.

Another method of controlling water hyacinth (*Eichhornia crassipes*) is integration of mechanical, chemical and biological methods because of the inadequacies of each method. Charudattan, (1986) developed a practical integrated water hyacinth control system based on a pathogen, *Cercospora rodmanii*; arthropods, including *Neochetina eichhorniae*; a moth, *Sameodes albiguttalis* and a mite, *Orthogalumna terebrantis* and 2 chemical herbicides; 2, 4-D and diquat.

He reported that the pathogen or the arthropods alone did not completely control water hyacinth, but that their combination was highly synergistic, yielding 99% control after 7 months.

The Problems Caused by Water Hyacinth Infestation: Water hyacinth can cause a variety of problems when its rapid mat-like proliferation covers areas of fresh water. Some of the common problems are listed below:

Hindrance to Water Transport: Access to harbours and docking areas can be seriously hindered by mats of water hyacinth. Canals and freshwater rivers can become impassable as they become clogged with densely intertwined carpets of the weed. The Ologe Lagoon and the entire Lagos Lagoon complex have been rendered almost impassable by this aquatic plant.

Clogging of Intakes of Irrigation, Hydropower and Water Supply Systems: Many large hydropower schemes are suffering from the effects of water hyacinth. The Owen Falls hydropower scheme at Jinja on Lake Victoria is a victim of the weeds rapid reproductive rates and an increasing amount of time and money has to be invested in clearing the weed to prevent it entering the turbine and causing damage and power interruptions.

Water hyacinth is now a major problem in some of the world's major dams-the Kariba dam which straddles the Zambia-Zimbabwe border on the Zambezi River and feeds Harare has pronounced infestations of the weed.

Blockage of Canals and Rivers Causing Flooding: If water hyacinth is allowed to grow uncontrolled, it can block canals and rivers, which could lead to flooding.

Micro-habitat for a Variety of Disease Vectors: The diseases associated with the presence of aquatic weeds in tropical developing countries are among those that cause the major public health problems: malaria, schistosomiasis and lymphatic filariasis. Some species of mosquito larvae thrive on the environment created by the presence of aquatic weeds, while the link between schistosomiasis (bilharzia) and aquatic weed presence is well known. Although, the statistical link is not well defined between the presence of aquatic weeds and malaria and schistosomiasis, it can be shown that the brughian type of filariasis (which is responsible for a minor share of lymphatic filariasis in South Asia) is entirely linked to the presence of aquatic weeds.

Increased Evapotranspiration: Various studies have been carried out to ascertain the relationship between aquatic plants and the rate of evapotranspiration compared with evaporation from an open-surfaced water body. Haider (1989) suggested that the rate of water loss due to evapotranspiration can be as much as 1.8 times that of evaporation from the same surface but free of plants. This has great implications where water is already scarce. It is estimated that the flow of water in the Nile could be reduced by up to one tenth due to increased losses in Lake Victoria from water hyacinth.

Problems Related to Fishing: Water hyacinth can present many problems for the fisherman. Access to sites becomes difficult when weed infestation is present, loss of fishing equipment often results when nets or lines become tangled in the root systems of the weed and the result of these problems is more often than not, a reduction in catch and subsequent loss of livelihood. Mailu (2001) reported that information from the Fisheries Department, Kenya indicated that there was a 28% increase in total annual fish catches between 1986-1991 and 1991-1997, from 133,097 to 169,890 tonnes.

There was an increase in all species of fish caught except *Oreochromis, Clarias* and *Mormyrus,* which showed declines of 14, 37 and 59%, respectively, over the same period. These declines may have been associated with the inability of fishermen to access the fishing grounds for these species because of water hyacinth infestation.

Generally therefore, as a result of water hyacinth infestation, accessibility to land and water has been hindered, resulting in reduced fish catches, especially of tilapia and mudfish which are found mainly along the shores. Fisherfolks, however, reported increased fish catches from suitable breeding grounds provided by water hyacinth e.g., tilapia, synodontis, protopterus and labeo (Mailu, 2001).

There is, however, need to clarify this conflicting information; in many more areas around the lake and in other parts of the world. A reduced fish catch would have an adverse effect on the quality of life of the communities around the lake and consequently affect sustainable development in the region.

Reduction of Biodiversity: Infestations of water hyacinth affect biodiversity. Dense mats of the weed covering the water surface lead to deoxygenation of the water, thus affecting all aquatic organisms. Death of water hyacinth mats may influence changes in the composition, distribution and diversity of aquatic organisms as follows:

- Displacement of hydrophytes and depressed algal biomass (Twongo *et al.*, 1995; Twongo and Balirwa, 1995)
- Increase in diversity and abundance of some taxa of macrofauna, especially at the borders of the weed mats (Wanda, 1997)
- Increase in the distribution and abundance of schistosome (bilharzia) snail vectors e.g., *Biomphalaria* sp. and *Bulinus* sp.
- Willoughby *et al.* (1993) reported that, based on studies on the

Ugandan shoreline of Lake Victoria, mats significantly depressed the diversity of fish species and fish biomass. It was subsequently demonstrated that fish diversity, particularly small taxa, increased along the edge of water hyacinth mats (Twongo and Balirwa, 1995)

Water Supply: Water supply to both villages and municipalities is affected by water hyacinth. In municipalities, water hyacinth interferes with the water intake points through blockage, which lowers the quantity of water pumped. In Kisumu, the municipality reports that the quantity of water supplied has dropped from 20,000 to 10,000 m^3 per day.

Water hyacinth infestations have been reported to lower the water quality in Kenya and Uganda (in terms of colour, pH, turbidity (suspended solids) of water) and hence increase the treatment costs.

Increased costs are associated with keeping the water intake points free of water hyacinth. For example, Kisumu Municipality employs 12 casuals per day, 6 drivers and 6 boat operators, while Homa Bay municipality engages 2 divers at a cost of 1000 Kenya shillings (Ksh) per day. In Homa Bay municipality, water hyacinth builds up 3 to 4 times in a week and it takes 3-4 h to remove it.

Uses of Water Hyacinth

Paper: The Mennonite Central Committee of Bangladesh has been experimenting with paper production from water hyacinth for some years. They have established two projects that make paper from water hyacinth stems. The water hyacinth fibre alone does not make a particularly good paper but when the fibre is blended with waste paper or jute, the result is good. The pulp is dosed with bleaching powder, calcium carbonate and sodium carbonate before being heated.

The first project is quite large with 120 producers involved in paper manufacture. The equipment for pulping is relatively sophisticated and the end product is of reasonable quality. The second project involves 25-30 people and uses a modified rice mill to produce pulp. The quality of the paper is low and is used for making folders, boxes, etc. Similar small-scale cottage industry papermaking projects have been successful in a number of countries, including the Philippines, Indonesia and India.

Fibre Board: Another application of water hyacinth is the production of fibre boards for a variety of end uses. The House and Building Research Institute in Dhaka has carried out experimental work on the production of fibre boards from water hyacinth fibre and other indigenous materials. They have developed a locally manufactured production plant for producing fibreboard for general-purpose use and also a bituminised board for use as a low cost roofing material.

The chopped water hyacinth stalks are reduced by boiling and then washed and beaten. The pulp is bleached and mixed with waste paper pulp and a filter agent such as china clay and the pH is balanced. The boards are floated in a vat on water and then finished in a hand press and hung to dry. The physical properties of the board are sufficiently good for use on indoor partition walls and ceilings.

Yarn and Rope: The fibre from the stems of the water hyacinth plant can be used to make rope. The stalk from the plant is shredded lengthways to expose the fibres and then left to dry for several days. The rope making process is similar to that of jute rope.

The finished rope is treated with sodium metabisulphite to prevent it from rotting. In Bangladesh, the rope is used by a local furniture manufacturer who winds the rope around a cane frame to produce an elegant finished product.

Basket Work: In the Philippines, water hyacinth is dried and used to make baskets and matting for domestic use. The key to a good product is to ensure that the stalks are properly dried before being used. If the stalks still contain moisture then this can cause the product to rot quite quickly. In India, water hyacinth is also used to produce similar goods for the tourist industry. Traditional basket making and weaving skills are used.

Charcoal Briquetting: This is an idea which has been proposed in Kenya to deal with the rapidly expanding carpets of water hyacinth which are evident on many parts of Lake Victoria. The proposal is to develop a suitable technology for the briquetting of charcoal dust from the pyrolysis of water hyacinth. It is suggested that a small-scale water hyacinth charcoal briquetting industry could have several beneficial aspects for the lakeside communities:

- Providing an alternative income
- Providing an alternative source of biomass
- Improvement of the lake shore environment through the removal of water hyacinth
- Improved access to the lake and less risk to maritime transport

- Reduced health risk associated with the presence of water hyacinth
- Alleviation of pressure on other biomass fuel sources, such as wood, thereby reducing deforestation and associated soil erosion.

The technical aspects are yet to be fully developed and tested but 7 main stages have been identified in the process of converting the plant into charcoal briquettes:

- Harvesting and collection of the plant
- Drying
- Collection and transport to the kiln
- Pyrolysis
- Mixing of the resultant dust with a binder
- Pressing into briquettes
- Marketing of briquettes.

Biogas Production: The possibility of converting water hyacinth to biogas has been an area of major interest for many years. Conversion of other organic matter, usually animal or human waste to biogas is a well established small and medium scale technology in a number of developing countries, notably in China and India. The process is one of anaerobic digestion which takes place in a reactor or digester (an air tight container usually sited below ground) and the usable product is methane gas which can be used as a fuel for cooking, lighting or for powering an engine to provide shaft power. The residue from the digestion process provides an organic fertilizer rich in nutrients.

The use of water hyacinth for digestion in a traditional digester presents some problems. Water hyacinth has very high water content and therefore harvesting effort yields a low reward in terms of organic matter for conversion to biogas.

The digester size has to be large compared with that of a traditional type due to the low gas production to plant volume ratio and this can in turn present problems for obtaining an airtight seal. Water hyacinth has to be pre-treated before entering the digester (macerated, chopped or beaten) to promote digestion and to remove air entrapped in the tissue of the plant which would cause it to float. To reduce the need for large volume digesters, high rate digestion techniques have been employed.

One such design has been tested in Bangladesh by a team from Warwick University, UK and the Housing and Building Research

Institute, Dhaka, Bangladesh. The design was for a small (8.3 cm) baffled reactor which was fed with juiced water hyacinth.

The throughflow of the reactor was 1.2 cubic metres per day. Some cow dung and rumen (taken from a cow's stomach) was added to the water hyacinth juice to promote digestion. Gas was produced in reasonable quantities but some problems were experienced with throughflow and further development is still required.

Other studies have been carried out, primarily in India with quantities of up to 4000 l of gas per tonne of semi dried water hyacinth being produced with a methane content of up to 64%. Most of the experiments have used a mixture of animal waste and water hyacinth. There is still no firm consensus on the design of an appropriate water hyacinth biogas digester.

Water Purification: Water hyacinth can be used to aid the process of water purification either for drinking water or for liquid effluent from sewage systems. In a drinking water treatment plant, water hyacinth has been used as part of the pretreatment purification step. Clean, healthy plants have been incorporated into water clarifiers and help with the removal of small flocs that remain after initial coagulation and floc removal or settling.

The result is a significant decrease in turbidity due to the removal of flocs and also slight reduction in organic matter in the water. In sewage systems, the root structures of water hyacinth (and other aquatic plants) provide a suitable environment for aerobic bacteria to function. Aerobic bacteria feed on nutrients and produce inorganic compounds which in turn provide food for the plants. The plants grow quickly and can be harvested to provide rich and valuable compost. Water hyacinth has also been used for the removal or reduction of nutrients, heavy metals, organic compounds and pathogens from water.

Animal Fodder: Studies have shown that the nutrients in water hyacinth are available to ruminants. In Southeast Asia, some non-ruminant animals are fed rations containing water hyacinth. In China, pig farmers boil chopped water hyacinth with vegetable waste, rice bran, copra cake and salt to make a suitable feed. In Malaysia, fresh water hyacinth is cooked with rice bran and fishmeal and mixed with copra meal as feed for pigs, ducks and pond fish.

Similar practices are much used in Indonesia, the Philippines and Thailand). The high water and mineral content mean that it is not suited to all animals. The use of water hyacinth for animal feed in developing countries could help solve some of the nutritional problems that exist in these countries. Animal feed is often in short supply and although humans cannot eat water hyacinth directly, they can feed it to cattle and other animals which can convert the nutrient into useful food products for human consumption.

Fertilisers: Water hyacinth can be used on the land either as a green manure or as compost. As a green manure, it can be either ploughed into the ground or used as mulch. The plant is ideal for composting. After removing the plant from the water, it can be left to dry for a few days before being mixed with ash, soil and some animal manure. Microbial decomposition breaks down the fats, lipids, proteins, sugars and starches.

The mixture can be left in piles to compost, the warmer climate of tropical countries accelerating the process and produces rich pathogen-free compost which can be applied directly to the soil.

The compost increases soil fertility and crop yield and generally improves the quality of the soil. Compost can be made on a large or small scale and is well suited to labour intensive, low capital production. In developing countries where inorganic fertilizer is expensive, it is an elegant solution to the problem of water hyacinth proliferation and also poor soil quality. In Sri Lanka, water hyacinth is mixed with organic municipal waste, ash and soil, composted and sold to local farmers and market gardeners.

Fish Feed: The Chinese grass carp is a fast growing fish which eats aquatic plants. It grows at a tremendous rate and reaches sizes of up to 32 kg). It is an edible fish with a tasty white meat. It will eat submerged or floating plants and also bank grasses. The fish can be used for weed control and will eat up to 18-40% of its own body weight in a single day). Other fish such as the tilapia, silver carp and the silver dollar fish are all aquatic and can be used to control aquatic weeds. The manatee or sea cow has also been suggested as another herbivore which could be used for aquatic weed control. Water hyacinth has also been used indirectly to feed fish. Dehydrated water hyacinth has been added to the diet of channel catfish fingerlings to increase their growth. It has also been noted that decay of water hyacinth after chemical control releases nutrients which promote the growth of phytoplankton with subsequent increases in fish yield.

Phytoremediation of Heavy Metals and Crude Oil-polluted Site: Several authors) have reported on the ability of water hyacinth to absorb heavy metals and rid the aquatic environment of these pollutants. The results are quite encouraging. As a matter of fact, reported on passive phytoremediation of heavy metal in Ologe Lagoon, Lagos, Nigeria by water hyacinth. They reported that water hyacinth that was not deliberately introduced into the lagoon to absorb heavy metals did so even when the concentration of the heavy metals in the water column was very small.

However, the study on phytoremediation of petroleum hydrocarbon by water hyacinth is still at the infancy. Few studies have been done on it which indicates that water hyacinth can absorb crude oil but more still needs to be done to fully establish the efficacy of water hyacinth in phytoremediation of petroleum hydrocarbon.

6

General Key in Aquatic Plants

Although this guide is primarily devoted to managing or reducing the amount of aquatic vegetation present in a body of water, the reader should be aware that aquatic plants have many important functions in aquatic systems. Just as on land, plants in water are natural and essential components of the environment, and without them, life as we know it could not exist. All plants, whether on land or in the water, utilize sunlight, carbon dioxide, and water to photosynthesize. The process of photo synthesis results in the production of new plant tissue (biomass) and oxygen.

In water, new plant biomass takes the form of either microscopic plants or larger plants called macrophytes. Why are these plants so important to the aquatic environment?

1. Microscopic plants (algae) form the base of the aquatic food chain. Another term for these plants is ìphytoplanktonî or plant plankton. They, in turn, are fed upon by zooplankton, or the microscopic animal plankton. Zooplankton are fed upon by small fish, small fish by larger fish, and so on up the food chain to humans and other top predators.
2. The macrophytes (larger algae and flowering plants) provide habitat and shelter for fish, fish food organisms, water fowl, and other wildlife. 3. Macrophytes provide food for insects, water fowl, and mammals such as muskrats and beaver. However, bass, bluegill, and catfish do not, as a rule, eat macrophytic vegetation.
4. All plants produce oxygen as they photo synthesize during the daylight hours. Photosynthesis is the major source of oxygen for aquatic animal life.

5. Rooted plants stabilize shorelines and bottom sediments; uprooting these plants can lead to shoreline erosion and increased turbidity in a body of water.
6. A diversity of aquatic plant life, both around and in the water, can add visual interest and beauty to a water body. One of the largest growing sectors of the home nursery industry is water gardening. Many people now recognize and appreciate the aesthetic and natural qualities of aquatic vegetation, whether in a backyard setting, around a retention pond in an urban area, or along the shoreline of a large lake. Because of these many benefits, some aquatic plant growth is always desirable. The goal of man agreement should never be to totally eliminate aquatic vegetation from a site.

Problems Caused by Excessive Aquatic Vegetation

Unfortunately, plant growth can get out of hand. Excessive plant growth is usually due to the fact that many of our bodies of water are shallow, which allows sunlight to penetrate to the bottom and support photosynthesis. Also, our bodies of water tend to be rich in nutrients such as nitrogen and phosphorus (substances that are needed by plants for growth). In some cases, exotic species of plants that are extremely aggressive have been introduced and have taken over large areas of aquatic habitat. Some of the problems caused by excessive aquatic plant growth are as follows:

1. Recreational activities such as swimming, fishing, and boating can be impaired and even prevented.
2. Excessive growths can lead to fish stunting and overpopulation. This occurs because the production of too much habitat prevents effective predation of small fish by larger fish.
3. Aquatic plant and algae growths can play a role in causing fish kills. This usually occurs because oxygen is taken out of the water. During the day, plants produce oxygen through photo synthesis; at night (as well as day), they consume oxygen through respiration. If plant growth is excessive, plants at night can use up most of the oxygen in the water. In fact, fish that are stressed for oxygen often die just before dawn when the oxygen content is lowest. Sometimes they can be ob served coming to the surface and gasping for air during the early morning hours. This is a sure sign that the fish are oxygen stressed and may die. Oxygen depletion also occurs when algae

and plants die and decompose. When plants die, photo synthetic production of oxygen ceases, and the bacteria and fungi, which break down (decompose) the decaying plant material, use up the oxygen in their

own respiration. Unfortunately, it is difficult to predict when a plant die-off will occur under natural conditions. Changes in water temperature, pro longed periods of cloud cover, shifts in wind pat terns, and depletion of nutrients are just some of the factors that may trigger a plant or algae die-off. Another cause of plant death and oxygen depletion is the treatment of too much vegetation with a herbicide.

Similarly, it is almost impossible to predict when or if fish are going to die because of vegetation related oxygen depletion. Fish kills typically occur in the summer or in the winter. Summer fish kills can be caused, as already described, by die-offs and de composition of algae and plant growth. Another cause can be due to a dense cover of algae or free-floating plants such as duckweed or water meal on the surface of a body of water. These surface growths reduce light penetration to the deeper waters which in turn inhibits photosynthesis and oxygen production. Anything that stirs or brings these deoxygenated waters to the surface (such as oxygen levels throughout the water column and cause a h kill. Even if fish are not immediately killed, pro longed exposure to low oxygen concentrations can lead to greater fish susceptibility to diseases and toxicants. Fish kills in winter occur when snow accumulates on ice cover. Light is blocked, thus preventing photosynthesis and oxygen production by any living plants or algae. Decomposition of plants that died in the fall causes further oxygen depletion. The fact that a site has undergone a winter fish kill may not even be recognized because the dead fish can decompose and disappear under the ice or as the ice breaks. It may not be until spring or early summer that people notice that fishing is not what it used to be. Other causes of fish kills include natural stresses on fish as they come out of the winter and into the spawning season, insecticide runoff, ammonia leakage from storage tanks, and diseases.

4. Aquatic weed growth provides quiet water areas ideal for mosquito breeding.
5. Certain algae can impart foul tastes and odors to the water.
6. Weeds impede water flow in drainage ditches, irrigation canals, and culverts and cause water to back up.

7. Deposition of weeds, as well as sediment and debris, can cause the gradual filling in of bodies of water.
8. Excessive weed growth can lower property values and decrease aesthetic appeal of a body of water.
9. Exotic aquatic plant species (e.g., Eurasian water milfoil, purple loosestrife) can invade and completely take over stands of native vegetation.

This process upsets the natural balance in an aquatic system and can have adverse effects on the animals that depend on the native vegetation for habitat and food. What are the aquatic plants that most frequently cause these problems? The following guide is intended to help you identify these plants. What is an aquatic plant? The definition used in this document is: 'Aquatic plants are those whose photosynthetically active parts are submersed permanently or at least during the growing season'. This is a restrictive definition and means that most of the plants are obligate aquatics and not seasonal or marginal emergents. Most emergent and wetland plants only flower and fruit after water levels have subsided and many have only their roots submersed in water. The list of such species is quite extensive. The list of obligate aquatic plants, found only in deep permanent water of lakes and rivers is much shorter, about 100 species. Generally the species listed here reflects the authors experience in collecting plants in lakes, those species which were regularly found are included and a few commonly encountered marginal or non-obligate species are included. There is minimal overlap with the plants treated by MacKensie as Wetland Plants. With few exceptions Juncaceae, Cyperaceae and Gramineae are treated as wetland plants, not aquatics.

There will never be universal agreement as to where to draw the line between aquatic plants and wetland plants; these are arbitrary man-made boundaries in a natural continuum. We have attempted to have minimal overlap but make sure no species were omitted in assigning species to the Aquatic or Wetland category. Obligate aquatic plants usually grow in lakes and ponds or in permanent rivers, streams and sloughs. Some may survive for a while when water levels drop and their habitats are exposed but most grow in water deep enough that they are not normally exposed. Sub-tidal marine species are also included.

Rheophytes are excluded. They grow in habitats which are periodically inundated for short periods, often by flood waters, but they are established under non-flood conditions. They are firmly attached to rocks or trees to prevent being washed away by the swift currents. Rheophytes are more prevalent in tropical torrents. In high latitudes and altitudes many normally terrestrial plants may be found partially submersed in cold water much of the year.

Bryophytes are not treated except for the floating *Riccia fluitans* and *Ricciocarpus natans*. These two species may be conspicuous in some habitats and are distinct from the remaining submersed or marginal aquatic bryophytes. The treatment of Charophytes is relatively cursory. The major emphasis is on the flowering vascular plants or 'Aquatic Macrophytes'.

Not all of the plants listed are native or established outside of cultivation in British Columbia. Some species are aquarium plants that are introduced from time to time but do not persist, or are in cultivation but do not survive for long when they escape; others are major invasive weeds in other parts of the world which have not yet been found in BC but which may be expected eventually. People introduce many aquatic plants, both inadvertently and deliberately, some of which do eventually become established.

This manual includes one species of rush, *Juncus supiniformis*, three members of the Cyperaceae, *Scirpus subterminalis*, *Scirpus lacustris* and *Dulichium arundinaceum* but no grasses. Most members of the Cyperaceae, Juncaceae and Poaceae are marginal or wetland species which are difficult for non-specialists to identify at the species level. Some species may be found in the water, especially during high water, which are not included in these keys. The grass genus *Glyceria* is often found in shallow, marginal waters.

Field observations indicate that much of the variation in obligate aquatic plants which is often given specific recognition in herbaria may not be valid in nature. Many 'herbarium species' do not reflect reality. Aquatic plants are genetically plastic and respond to changing conditions with a tremendous amount of morphological variability. These morphological variants are responses to variable environmental conditions rather than indications of genetic distinctions at the species level. The species concept preferred by the author is a pragmatic 'lumping' approach, which may better reflect these field observations. However, in the interests of uniformity, the species concepts used in the four volume series 'The Vascular Plants of British Columbia' (ISSN 0843-6452) has been used in these keys. Occasional notes may be found to indicate the authors preference for another treatment.

The following pages include a list of all the species covered by these keys, a key to groups of aquatics based on their ecology, keys to each of the ecological groups, a general key to the aquatic plants of BC including a key to the aquatic plants with finely dissected submersed leaves, keys to the families and genera identified in the general key, a set of brief notes on the distribution, abundance and habitats of each species of aquatic plant, and references which may be useful in studying some aspects of the aquatic plants of BC. Illustrations of the types of dissected underwater leaves which may be found are included. Synonyms are used very sparingly and only for species which have other long-established and well known names. They are given, after the distribution and habitat data, in [brackets]. More synonyms may be found in 'The Vascular Plants of British Columbia' which also contains extensive references to the BC flora. A virtually complete list of common names gleaned from the world-wide literature is listed. There are many common names for widespread weedy species which grow in many different countries. Where more than one common name is given the preferred name for British Columbia is given first and the remainder follow in alphabetical order. Complex common names are often found in the literature and may be separate words, compound words or hyphenated words, they may also have capital or lowercase initial letters for each word. This can lead to a great many unique combinations. These combinations have been reduced to one multiple-word, non-hyphenated name with the first letter of the first word capitalized and all the rest of the words in lowercase.

The distribution limits of species are presented using vegetation zones as defined in 'The Vascular Plants of British Columbia'. The montane zone includes all continuous forests in BC except for the coastal lowlands and some islands included in the lowland zone. The subalpine zone is defined as that area above the montane zone and below the upper limit of conifers growing as an upright tree. This is represented by a meadow and tree-clump complex in the south and shrub Salix and scattered trees in the north. The alpine is above the subalpine where trees occur only as krummholz and vegetation is of tundra form. Steppe vegetation occurs in the interior and includes sagebrush or grasslands. The various keys are not mutually exclusive and may be used together to help restrict the choices or confirm the identification of a species. Using the Ecological key may limit the number of species one has to choose from in the genus or species keys, or in the dissected leaf key. Similarly if the plant obviously has dissected underwater leaves then using the dissected leaf key is more efficient than using the general key. The Ecological key may result in only a few choices which can then be readily compared with the illustrations for identification.

Some of the keys are in hierarchical sets. The Ecological key restricts you to a group of plants and the subsequent species key identifies the individual species. The key to families restricts you to genera and species if you know the family, and the key to genera identifies species if you know the genus. If you already know the family or genus you can go directly to the key that is appropriate to the level of your knowledge of the plant in question, and not have to start each time with the general key to all the aquatic plants. Using several different keys which should include the plant in question is a good check on your ability to use the keys and on the usefulness of the keys. You should get the same identification each time; if you do not perhaps you have made some assumptions about the plant which are not true or have taken the wrong fork in one of the keys. It is also possible that one of the keys can not handle some unanticipated variation in the specimen. All identifications made by keys should be checked with descriptions and illustrations, or better yet with herbarium specimens, to verify the identification. You may have a plant which is not included in the keys. The less experience you have with the species and the keys, the more important this verification becomes.

The general key to the aquatic plants of BC initially separates aquatic plants into 7 parts, for convenience and ease of use; this initial separation is based on ecology. Part 1 identifies herbaceous, fresh water plants which float freely on, or just under, the surface of the water. Part 2 identifies fully submersed, herbaceous, fresh water plants which are rooted or attached and have at least some finely dissected leaves. Part 3 identifies fully submersed, herbaceous, fresh water plants which are rooted or attached and do not have finely dissected leaves. Part 4 identifies herbaceous, fresh water plants which are rooted and emerge above the surface of the water but have cauline leaves that are opposite, whorled or clustered (more than one at each node). Part 5 identifies herbaceous fresh water plants which are rooted and emerge above the surface of the water but have cauline leaves that are alternate (only one at each node). Part 6 identifies herbaceous fresh water plants which are rooted and emerge above the surface of the water but have leaves in basal clusters as opposed to cauline leaves. Part 7 identifies fully submersed marine or brackish plants. With experience, or if you know which type of plant you have, you can go directly to the correct key. Marginal emergent shrubs such as *Potentilla palustris* are ommitted and dealt with as wetland species.

Ecological keys group plants without regard to the species and genus. They describe where and under what conditions the plants grow, in relation to the water level, and with regard to their morphology, which is often shaped by their habitat. Some are submersed, some float on the surface and some are emergent. A number of ecological classification systems have been proposed; all are useful for specific purposes and all have their limitations.

Some of these patterns are quite distinctive and can be readily given a name: *Myriophyllum* species are all pinnate and *Ceratophyllum* species dichotomous, but most of the other types are more complex and not readily given a simple name.

The Aquatic Plants of British Columbia

- *Alisma gramineum* J. G. Gmel.
- *Alisma plantago-aquatica* L.
- *Azolla caroliniana* Willd.
- *Azolla filiculoides* Lam.
- *Azolla mexicana* Presl.
- *Brasenia schreberi* Gmel.
- {*Cabomba caroliniana*} Gray
- *Calla palustris* L.
- *Callitriche anceps* Fern.
- *Callitriche hermaphroditica* L.
- *Callitriche heterophylla* Pursh
- *Callitriche stagnalis* Scop.
- *Callitriche verna* L.
- *Caltha natans* Pallas
- *Caltha palustris* L.
- *Ceratophyllum demersum* L.
- *Ceratophyllum echinatum* Gray
- *Chara braunii* Gm.
- *Chara canescens* Desv. and Lois.
- *Chara globularis* Thuill.
- *Chara vulgaris* L.
- *Crassula aquatica* (L.) Schoenl.
- *Dulichium arundinaceum* (L.) Britt.
- *Egeria densa* Planch.

- {*Eichhornia crassipes*} (Mart.) Solms
- *Elatine rubella* Rydb.
- *Elodea canadensis* Rich.
- {*Elodea longivaginata*} St. John
- *Elodea nuttallii* (Planch.) St. John
- *Equisetum fluviatile* L. em Ehrh.
- *Equisetum palustre* L.
- *Gratiola ebracteata* Bentham
- *Gratiola neglecta* Torr.
- *Heteranthera dubia* (Jacq.) Macmill.
- *Hippuris vulgaris* L.
- {*Hydrilla verticillata*} (L.) Royle
- *Hydrocotyle ranunculoides* L. f.
- *Hydrocotyle verticillata* Thunb.
- *Isoetes bolanderi* Engelm.
- *Isoetes echinospora* Dur.
- *Isoetes howellii* Engelm.
- *Isoetes maritima* Underw.
- *Isoetes nuttallii* A. Br.
- *Isoetes occidentalis* Henderson
- *Isoetes truncata* (A. A. Eaton) Clute
- *Juncus supiniformis* Engelm.
- {*Lagarosiphon major*} Ridley
- *Lemna minor* L.
- *Lemna trisulca* L.
- *Lilaea scilloides* (Poir.) Hauman
- *Lilaeopsis occidentalis* Coult. and Rose
- {*Limnobium laevigatum*} (Humb. and Bonpl.) Heine
- {*Limnobium spongia*} (Bosc) Steud.
- {*Limnophila sessiliflora*} Blume
- *Limosella aquatica* L.
- *Lobelia dortmanna* L.
- *Ludwigia palustris* (L.) Ell.
- *Lysimachia thyrsiflora* L.

- *Marsilea vestita* Hook. and Grev.
- *Megalodonta beckii* (Torr. ex Spreng.) Greene
- *Menyanthes trifoliata* L.
- *Montia fontana* L.
- {*Myriophyllum aquaticum*} (Vell.) Verd.
- *Myriophyllum farwellii* Morong
- {*Myriophyllum heterophyllum*} Michx.
- *Myriophyllum hippuroides* Nutt.
- *Myriophyllum pinnatum* (Walt.) B. S. P.
- *Myriophyllum quitense* H. B. K.
- *Myriophyllum sibiricum* Kom.
- *Myriophyllum spicatum* L.
- *Myriophyllum ussuriense* (Regel) Maxim.
- *Myriophyllum verticillatum* L.
- *Najas flexilis* (Willd.) R. and S.
- *Nasturtium officinale* R. Br. in W. Ait
- *Nitella acuminata* A. Br.
- *Nitella clavata* Kutz.
- *Nitella flexilis* (L.) Ag.
- *Nitella furcata* (Roxb.) Ag.
- *Nitella gracilis* (Sm.) Ag.
- *Nitella tenuissima* (Desv.) Kutz.
- *Nuphar polysepalum* Engelm.
- *Nuphar variegatum* Engelm. ex Durand
- *Nymphaea alba* L.
- *Nymphaea leibergii* Morong
- *Nymphaea mexicana* Zucc.
- *Nymphaea odorata* Ait.
- *Nymphaea tetragona* Georgi
- {*Nymphoides aquatica*} (Gmelin) O. Kuntze
- {*Nymphoides cordatum*} (Ell.) Fernald
- {*Nymphoides peltata*} (Gmelin) O. Kuntze
- *Phyllospadix scouleri* Hooker
- *Phyllospadix serrulatus* Rupr.

- *Phyllospadix torreyi* S. Wats.
- {*Pilularia americana*} R. Br.
- {*Pistia stratiotes*} L.
- *Polygonum amphibium* L.
- *Polygonum hyropiper* L.
- *Polygonum hydropiperoides* Michx.
- *Polygonum lapathifolium* L.
- {*Pontederia cordata*} Lour.
- *Potamogeton alpinus* Balbis
- *Potamogeton amplifolius* Tuckerman
- *Potamogeton berchtoldii* Fieb. in Bercht.
- *Potamogeton crispus* L.
- *Potamogeton epihydrus* Raf.
- *Potamogeton filiformis* Pers.
- *Potamogeton foliosus* Raf.
- *Potamogeton friesii* Rupr.
- *Potamogeton gramineus* L.
- *Potamogeton illinoensis* Morong
- *Potamogeton natans* L.
- *Potamogeton nodosus* Poir.
- *Potamogeton oakesianus* Robbins
- *Potamogeton obtusifolius* Mertens and Koch
- *Potamogeton pectinatus* L.
- *Potamogeton perfoliatus* L.
- *Potamogeton praelongus* Wulf.
- *Potamogeton pusillus* L.
- *Potamogeton richardsonii* (A. Benn.) Rydb.
- *Potamogeton robbinsii* Oakes
- *Potamogeton strictifolius* Bennett
- *Potamogeton vaginatus* Turcz.
- *Potamogeton zosteriformis* Fern.
- *Ranunculus aquatilis* L.
- *Ranunculus circinatus* Sibth.
- *Ranunculus cymbalaria* Pursh

- *Ranunculus flabellaris* Raf.
- *Ranunculus flammula* L.
- *Ranunculus gmelinii* DC.
- *Ranunculus hyperboreus* Rottb.
- *Ranunculus lobbii* (Hiern) A. Gray
- *Ranunculus sceleratus* L.
- *Riccia fluitans* L.
- *Ricciocarpus natans* (L.) Corda
- *Ruppia maritima* L.
- *Sagittaria cuneata* Sheld.
- *Sagittaria latifolia* Willd.
- *Salvinia* Seguier
- *Scheuchzeria palustris* L.
- *Scirpus lacustris* L.
- *Scirpus subterminalis* Torr.
- *Sium sauve* Walt.
- *Sparganium angustifolium* Mich.
- *Sparganium emersum* Rehm.
- *Sparganium eurycarpum* Engelm.
- *Sparganium fluctuans* (Morong) Robbins.
- *Sparganium glomeratum* Laest. ex Beurl.
- *Sparganium hyperboreum* Laest. ex Beurl.
- *Sparganium natans* L.
- *Spirodela polyrhiza* (L.) Schleid.
- *Subularia aquatica* L.
- *Tolypella intricata* (Trent.) Leonh.
- *{Trapa natans}* L.
- *Utricularia gibba* L.
- *Utricularia intermedia* Hayne
- *Utricularia minor* L.
- *Utricularia vulgaris* L.
- *Vallisneria americana* Michx.
- *Veronica anagallis-aquatica* L.
- *Veronica beccabunga* L.

- *Veronica catenata* Pennel
- *Veronica scutellata* L.
- *Wolffia borealis* (Engelm. ex Hegelmaier) Landolt and Wildi
- *Wolffia columbiana* Karst.
- *Wolffiella floridana* (J. D. Smith) Thompson
- *Zannichellia palustris* L.
- *Zostera japonica* Ascher and Grabn.
- *Zostera marina* L.

General Key to the Aquatic Plants of British Columbia

- plants of lakes, ponds, rivers and other permanent freshwater habitats, all herbaceous, rooted emergents, floating or submersed.
- plants all floating freely on the surface of the water, or just under the surface, not rooted or attached, except sometimes when stranded. General Key: Aquatic Plants of British Columbia: Part 1.
- plants submersed or emergent, generally rooted or attached, not freely floating on or at the surface.
- plants fully submersed, leaves may float on the surface and flowers may be emergent, stems and petioles remain on or under the water.
- plants with at least some, and often all, the underwater leaves finely dissected, and with at most a few floating leaves. General Key: Aquatic Plants of British Columbia: Part 2.
- plants lacking finely dissected submersed leaves, leaves may all float on the surface, may all be fully submersed or be some combination of both. General Key: Aquatic Plants of British Columbia: Part 3.
- plants rooted in the sediment but emergent, much, if not most or all, of the stem is emergent for most or all of the year.
- leaves cauline, more or less evenly distributed along the stem.
- leaves opposite or whorled, in groups, or clusters of several leaves. General Key: Aquatic Plants of British Columbia: Part 4.
- leaves alternate, only one at a node. General Key: Aquatic Plants of British Columbia: Part 5.
- leaves in basal clusters or bunches, not evenly distributed along an elongate stem. General Key: Aquatic Plants of British Columbia: Part 6.

- plants of marine or brackish lagoon habitats. General Key: Aquatic Plants of British Columbia: Part 7.

Part 1: Floating Plants

A microscope is needed for positive identification of *Azolla* species.

- Floating on the surface of the water.
- Large plants with distinct roots, stems and leaves; may have an obvious swollen part used as a floatation organ.
- most leaves are erect and emergent, in compact rosettes or distributed along the stem.
- petioles inflated to serve as floats, leaf bases cordate or abruptly narrowed to the petiole, flowers large, showy and purple, leaves glabrous and margins entire, rounded in outline, parallel veined. - {*Eichhornia crassipes*}
- leaves sessile or sub-sessile, no inflated petiole, leaves pubescent.
- plants with all leaves in a compact rosette arising from a common point, leaves obovate and apically notched, flowers in short inflorescences, shorter than the greyish-green and densely pubescent large leaves, flowering plants. - {*Pistia stratiotes*}
- plants with small hairy leaves distributed along the stems, ferns. - {*Salvinia*}
- leaves lie flat on the surface of the water, leaves long petiolate, arising from a common point.
- leaf bases cuneate to truncate, flowers short-petiolate and inconspicuous, leaves rhombic and angular, apical margins at least toothed, leaves not thick and spongy but the petiole has a small swollen section, at the leaf junction, for a float. - {*Trapa natans*}
- leaves round to heart-shaped with a basal notch, leaf blades spongy, petioles without a swollen section. - {*Limnobium*}
- leaves with a thick spongy layer, convex on the dorsal surface, short petiole and closely clustered unmarked leaves. - {*Limnobium laevigatum*}
- leaves with a thin spongy layer, flat on the dorsal surface, petioles longer than the remote leaf blades, leaves with red-brown markings on the dorsal surface. - {*Limnobium spongia*}
- Small thalloid plants with roots below, closely appressed to the surface of the water.

- thalli not repeatedly and regularly forked, elongate to orbicular, flat or globose, flowering plants (duckweeds) and ferns (*Azolla*).
- ferns (*Azolla*), stem branched above every third leaf, leaves minute, alternate, sessile, bi-lobed, one lobe submersed as a float, the other emergent and colonized by blue-green algae. - *Azolla*
- glochidia with several (usually 3 or more) septa, leaves at least 0.7 mm long, plant 1 to 3 cm in diameter, submersed leaf-lobes much larger than the upper lobes, glochidia unbranched. - *Azolla mexicana*
- glochidia with 1 (2) or no septa.
- glochidia without any septa, leaves papillose, oblong to ovate, about 1 mm long, megaspores coarsely roughened, plants 1 to 10 cm long, branches numerous and open, lower submersed leaf-lobe about as large as the upper lobe, glochidia unbranched. - *Azolla filiculoides*
- glochidia with 1 or 2 apical septa, leaves nearly smooth, sub-orbicular, about 0.5 mm long, megaspores finely roughened, plants usually less than 3 cm long, few branches and crowded, lower submersed leaf-lobe glabrous, larger and paler than the upper lobe, glochidia may be branched. - *Azolla caroliniana*
- flowering plants (duckweeds), no elongate stem, leaves solitary or in clusters, not bi-lobed, no leaves specially adapted as floats, no algae colonies. - Lemnaceae
- one or more roots and nerves on each thallus (frond).
- each thallus bears two or more clustered roots from the base and 4 to 12 nerves. - *Spirodela polyrhiza*
- each thallus bears one root at the base and 1 to 3 nerves. - *Lemna*
- fronds oblong to lanceolate, 6 to 12 mm long and connected in small groups by stalks of the same length, matted, generally submersed, colonies. - *Lemna trisulca*
- fronds oval to round, less than 6 mm long, no stalks, solitary or in small attached groups, floating on the surface, usually 1 nerved, 2 to 4 mm long. - *Lemna minor*
- no roots or nerves on the thalli.
- thalli sub-globose to oval, up to 1 mm long. - >b>*Wolffia*
- plants floating just below the surface, sub-globose, upper surface rounded, green, not puncticulate, 0.5 to 1.0 mm long. - *Wolffia columbiana*

- plants floating on the surface, ellipsoidal, upper surface flattish, white- or brown-punctilate, 0.5 to 1.2 mm long and about 1/2 as wide. - *Wolffia borealis*
- thalli long and narrow, sickle-shaped, several mm long but very narrow, flattened, submersed except at the base and aggregated into clusters. - {*Wolffiella floridana*}
- thalli repeatedly and regularly forked, bryophytes (liverworts).
- thalli broad, 2 to 3 times forked, tinged with purple, and fringed with tongue-shaped toothed scales. - *Ricciocarpus natans*
- thalli narrow and elongate, repeatedly dichotomously forked, not purple tinged nor fringed with scales. - *Riccia fluitans*
- Floating beneath the surface of the water.
- Found just under the surface tension of the water or free in the mid-water zone; often found tangled in other submersed or marginal vegetation.
- plants with roots, stems and dissected leaves.
- plants with bladders on the alternate leaves. - *Utricularia*
- leaves divided into ultimately terete or threadlike segments.
- leaves divided into fewer than 5 final threadlike segments, leaf margins glabrous, bladders scarce on a small delicate plant usually floating at the surface or entangled in other rooted plants. - *Utricularia gibba*
- leaves 'pinnatifid', more than 20 terete final segments, hairy leaf margins, many bladders on the ordinary leaves, a robust plant to several meters long, usually lying on the sediment surface. - *Utricularia vulgaris*
- leaves divided di- or tri-chotomously into ultimately flattened segments, ordinary leaves generally with a few bladders, leaf margins glabrous, the terminal leaf segments are acuminate. - *Utricularia minor*
- plants without bladders on the whorled leaves. - *Ceratophyllum*
- leaf segments sub-capillary, mostly entire, delicate and light green, in deeper water and not surfacing, achene with 3 to 5 lateral spines on each side, not a 'weedy' species. - *Ceratophyllum echinatum*
- leaf segments capillary to linear and flattened, serrate to coarsely toothed, plant usually coarse and robust, dark green to almost black, usually surfacing, achene without lateral spines,

2 basal spines and 1 terminal spine only, a very 'weedy' species in eutrophic waters. - *Ceratophyllum demersum*

- plants reduced to thalli with no stems or leaves, roots may be present. - Lemnaceae
- one or more roots and nerves on each thallus, fronds oblong to lanceolate, 6 to 12 mm long and connected in small groups by stalks of the same length, matted, generally submersed, colonies. - *Lemna trisulca*
- no roots or nerves on the thalli.
- thalli sub-globose to oval, up to 1 mm long. - *Wolffia*
- plants floating just below the surface, sub-globose, upper surface rounded, green, not puncticulate, 0.5 to 1.0 mm long. - *Wolffia columbiana*
- plants floating on the surface, ellipsoidal, upper surface flattish, white-or brown-punctilate, 0.5 to 1.2 mm long and about 1/2 as wide. - *Wolffia borealis*
- thalli long and narrow, sickle-shaped, several mm long but very narrow, flattened, submersed except at the base and aggregated into clusters. - {*Wolffiella floridana*}
- Found lying on or just above the sediment surface.
- plants reduced to thalli with no stems or leaves, 1 root, fronds oblong to lanceolate, 6 to 12 mm long and connected in small groups by stalks of the same length, matted, generally submersed, colonies. - *Lemna trisulca*
- plants with roots, stems and dissected leaves bearing bladders. - *Utricularia*
- leaves divided into ultimately terete or threadlike segments.
- leaves divided into fewer than 5 final threadlike segments, leaf margins glabrous, bladders scarce on a small delicate plant usually floating near the surface or entangled in other rooted plants. - *Utricularia gibba*
- leaves 'pinnatifid', more than 20 terete final segments, hairy leaf margins, many bladders on the ordinary leaves, a robust plant to several meters long, usually lying on the sediment surface. - *Utricularia vulgaris*
- leaves divided di- or tri-chotomously into ultimately flattened segments, ordinary leaves generally with a few bladders, leaf margins glabrous, the terminal leaf segments are acuminate. - *Utricularia minor*

Part 2: Plants With Finely Dissected Submersed Leaves

- Submersed leaves pinnately divided, 1 to 5 cm long.
- submersed leaves bi-pinnate or simply pinnate with at least the lowest segments forked, one fork pinnate and the other simple. (aquarium plant) - {*Limnophila sessiliflora*}
- submersed leaves simply pinnate (native and introduced plants widely used in aquaria and outdoor pools). - *Myriophyllum*
- flowers in the axils of cauline submersed leaves.
- leaves in whorls of 3 to 4 or scattered, fewer than 10 leaf segments on each side of the rachis, monoecious, 4 stamens, plants often reddish, fully submersed. - *Myriophyllum farwellii*
- leaves in whorls of 4 to 6, more than 10 leaf segments on each side of the rachis, dioecious, 8 stamens, plants pale yellowish-green, the apical portion of the stem often sprawled over the surface of the water or on the adjacent shore. - {*Myriophyllum aquaticum*}
- flowers in the axils of bracts on emergent, terminal spikes.
- four stamens, 4 to 6 leaves per whorl, bracts conspicuous.
- floral bracts delicate, deeply incised to serrate, spike short and delicate, known only from sloughs in the Fraser Valley. - *Myriophyllum hippuroides* and *Myriophyllum pinnatum*
- floral bracts ovate and toothed, spike long, robust and inflated introduced in several park and garden ponds of the south-west. - {*Myriophyllum heterophyllum*}
- eight stamens, 3 to 5 leaves per whorl, floral bracts various.
- floral bracts smaller than the flowers, inconspicuous, and nearly entire (the lowest few may be larger and pinnate but the upper ones are small), leaf whorls in the central portion of the stem are over 1 cm apart and not crowded, monoecious.
- no turions, rhizomatous, 10 to 16 leaf divisions less than 2 mm apart, leaves make right or obtuse angles with the stem, leaf tips 'squared', all leaf segments straight and all of nearly the same length. - *Myriophyllum spicatum*
- turions, not rhizomatous, 6 to 12 leaf segments over 2.5 mm apart, leaves make acute angles with the stem, leaf tips 'acute', basal leaf segments curved and much longer than the apical ones. - *Myriophyllum sibiricum* [*Myriophyllum exalbescens*]
- floral bracts usually longer than the flowers and rarely entire, leaf whorl spacing varies, monoecious or dioecious.

- dioecious, flowering plants found only on exposed mud banks when water levels drop in summer, female bracts and leaves entire to scarcely and irregularly divided, male bracts and leaves entire to pectinate-pinnate, submersed leaves scattered and irregular. - *Myriophyllum ussuriense*
- dioecious or monoecious, flowering plants found in water, bracts large and conspicuous, variable.
- floral bracts pinnate to pectinate, light green, leaves often crowded on the stem (short internodes) and delicate, usually with more than 10 leaf divisions. - *Myriophyllum verticillatum*
- floral bracts pectinately parted below becoming dentate in the middle and almost entire above, reddish, leaves well spaced on the stem (long internodes) and robust, leaves generally with fewer than10 leaf divisions. - *Myriophyllum quitense*
- submersed leaves palmately or di- to tri-chotomously divided but not pinnate, length is variable.
- leaves sessile, in whorls of 5 to 12, up to 2 cm long, toothed or serrate, plants are fully submersed and have no roots. - *Ceratophyllum*
- leaf segments sub-capillary, mostly entire, delicate and light green, deeper water and not surfacing, achene with 3 to 5 lateral spines on each side. - *Ceratophyllum echinatum*
- leaf segments capillary to linear and flattened, serrate to coarsely toothed, usually coarse and robust, dark green to almost black, usually surfacing, achene without lateral spines, 2 basal spines and 1 terminal spine only. - *Ceratophyllum demersum*
- leaves petiolate, in whorls of 3, decussate, entire, plants rooted. - {*Cabomba caroliniana*}
- submersed leaves alternate or opposite, up to 8 cm long, toothed or entire, plants submersed or emergent and with or without roots.
- submersed leaves opposite, 2 to 6 cm long, many times divided into ultimately filiform segments.
- leaves 2 to 4 cm long, sessile, many times dichotomously divided, the leaf segments remain filiform. - *Megalodonta beckii*
- leaves up to 6 cm long, petiolate, many times palmately divided, leaflet tips are flattened and spathulate. - {*Cabomba caroliniana*}

- submersed leaves alternate, 0.3 to 8 cm long, few to many times divided into flat filiform segments or threadlike.
- leaves with bladders, sessile, petiolar base of leaves not swollen. - *Utricularia*
- leaves divided into ultimately terete or threadlike segments.
- submersed leaves in whorls.
- leaves divided into fewer than 5 final threadlike segments, leaf margins glabrous, bladders scarce on a small delicate plant usually floating at the surface or entangled in other rooted plants. - *Utricularia gibba*
- leaves 'pinnatifid', more than 20 terete final segments, hairy leaf margins, many bladders on the ordinary leaves, a robust plant up to several meters long and usually lying on the sediment surface. - *Utricularia vulgaris*
- leaves divided di- or tri-chotomously into ultimately flattened segments.
- ordinary leaves generally with a few bladders, leaf margins glabrous, the terminal leaf segments are acuminate. - *Utricularia minor*
- ordinary leaves rarely with bladders, bladders are found on separate subterranean branches, leaf margins hairy, the ultimate leaf segment tips are awned. - *Utricularia intermedia*
- no bladders on the sessile or petiolate leaves.
- a swollen float at the top of the petiole next to the blade in the floating leaves which are rhomboid, flowers white, perianth 4-merous, fruit a unique, long-stalked nut. - {*Trapa natans*}
- petiolar base of leaves is swollen, flowers white or yellow, fruit of 1-seeded nutlets, perianth 5-merous, emergent or floating leaves various but not rhomboid. - *Ranunculus*
- flowers white, submersed leaves dissected into ultimately filiform segments.
- Two to 7 glabrous, beakless achenes, plants and receptacle glabrous, pedicels in the axils of ternately lobed floating leaves, leaves 2 to 3 times divided into 8 to 12 segments. - *Ranunculus lobbii*
- Ten to 80 glabrous or hirsute achenes, plants and receptacles hirsute.

- Thirty to 80 achenes, flowers may have yellow bases, leaf blades sessile on the stipular base. - *Ranunculus circinatus*
- Ten to 25 achenes, flowers all white, leaf blades petiolate. - *Ranunculus aquatilis*
- flowers yellow, submersed leaves, if present, simple to ternately divided, leaves all or in part lobed, parted or ternately dissected to filiform segments, achenes pubescent.
- annuals, erect, not nodally rooting, achene beaks inconspicuous. - *Ranunculus sceleratus*
- perennials, floating or reclining, nodally rooting, achene beaks conspicuous.
- leaves deeply 3 parted with narrow lobes distally acute, achenes not corky margined, achene beaks 1/4 the length of the achenes. - *Ranunculus gmelinii*
- leaves, at least the submersed ones, 3 to 5 times ternately dissected into filiform segments less than 2 mm wide, achenes corky margined. - *Ranunculus flabellaris*

Part 3: Submersed Plants Without Finely Dissected Leaves

Section 1: Non-flowering Plants (Algae, Ferns and Quillworts-*Isoetes*):

A compound or dissecting microscope is required to positively identify species in *Isoetes*; magnification over 10x is needed to examine the megaspore surface.

- algae, plant body composed of 1-celled internodes, which may be covered with a multi-cellular cortex, and with branchlets at the nodes, no leaves, fully submersed, often foul smelling and encrusted with marl, spores reddish in spherical, axillary clusters. - Characeae
- branchlets simple, not further branched or segmented, coronula 5-celled. - *Chara*
- stem not corticate, stipulodes in 1 tier, and alternating with the branchlets, branchlets tipped with a corona, no spine cells. - *Chara braunii*
- stems, and at least basal branchlet segment, corticate.
- spine cells in fascicles, axial cortex haplostichous, 1 corticate, stipulodes in 2 tiers, spiny looking, found in saline or brackish waters. - *Chara canescens*

- spine cells solitary or geminate, axial cortex diplostichous or triplostichous, 2 or 3 corticate.
- axial cortex diplostichous, stipulodes in 2 tiers, spine cells absent or small, rarely geminate, quite variable and widespread in hard-water lakes, smooth but marl encrusted. - *Chara vulgaris*
- axial cortex triplostichous, stipulodes usually in 2 tiers but may be in 1 tier, obscure or absent, spine cells usually absent or obscure but may be clustered, variable and widespread in hard-water and soft-water lakes, smooth appearance. - *Chara globularis*
- branchlets compound, re-branched and/or segmented.
- branchlets simple but segmented like a string of sausages, end cells reduced and acute, heads coarse and like a bird's nest. - *Tolypella intricata*
- branchlets compound, re-branched, may be segmented or apparently so. - *Nitella*
- heteroclemous, 2 kinds of branchlets in a whorl, 1-forked alternating with 1-celled, compact, branchlets swollen. - *Nitella clavata*
- homeoclemous, all branchlets in a whorl the same, generally 1-forked.
- fertile heads small and densely compact, no mucous, generally no terminal dactyls on the sterile branchlets. - *Nitella acuminata*
- fertile heads absent or loose and not compact.
- dactyl apices not long acuminate, may be acute, blunt or apiculate, (a small variety may have 2-forked branchlets). - *Nitella flexilis*
- dactyl apices are long acuminate, gametangia on branchlets, generally not marl covered. - *Nitella acuminata*
- the end segment of the branchlets is 2-or-more-celled, the end cells are very much smaller than the penultimate cell, branchlets furcate, heads terminal, branchlets homeoclemous.
- robust plants, over 15 cm high, axes over 600 microns in diameter, dactyls may be small. - *Nitella furcata*
- smaller plants, seldom over 15 cm tall, axes 200 to 450 microns in diameter, dactyls uniform.
- lowest branchlet node fertile, dactyls 2- to 3-celled, heads may have mucous, oospore membrane granular or felt-like. - *Nitella gracilis*

- lowest branchlet node sterile, dactyls 2-celled, heads without mucous, oospore membrane reticulate. - *Nitella tenuissima*
- Quillworts or ferns, plant body composed of a multi-cellular stem with leaves.
- leaves circinate in the bud ('fiddleheads'), ferns with hairy sporocarps at the bases of the leaves, leaves arise in clusters from the nodes of a creeping, horizontal stem. - Marsiliaceae
- leaves with a slender petiole and a 4-foliate blade (like a 4-leaf clover). - *Marsilea vestita*
- leaves are filiform, without an expanded blade. - *Pilularia americana*
- leaves not circinate, quillworts, no hairy sporocarps, sessile leaves apically filiform with a wider, clasping base, arising in a tight spiral from the apex of a short, compact 2- or 3-lobed corm-like structure, sporangia on the inner face of the leaf base. - *Isoetes*
- corms 3-lobed, plants primarily terrestrial, grow in grassy, ephemeral pools which are wet in winter and dry in summer. - *Isoetes nuttallii*
- corms 2-lobed, plants primarily of lake bottoms but sometimes exposed in summer due to water level fluctuations.
- megaspores spiny.
- megaspores vary greatly in size and are often aborted, spines blunt and dense. (this is probably a hybrid of *Isoetes echinospora* and *Isoetes maritima*). - *Isoetes truncata*
- megaspores uniform in size, spines sharp, not crowded.
- plants flaccid, megaspore spines elongate with pointed tips, uniform in size, microspores smooth. - *Isoetes echinospora*
- plants rigid, megaspore spines stubby and blunt, shorter and denser along the ridges, microspores spiny. - *Isoetes maritima*
- megaspores smooth or with ridges but not spiny.
- lowermost leaves in 2 ranks, megaspores over 0.5 mm in diameter, microspores rugose and over 36μm long. - *Isoetes occidentalis*
- lowermost leaves spirally arranged, megaspores less than 0.5 mm in diameter, microspore spiny and less than 30μm long.
- base of the leaves blackened, hyaline wing margins of the leaves extending 1 to 5 cm above the sporangium, sporangia brown spotted. - *Isoetes howellii*

- base of the leaves green, hyaline wing margins of the leaves not extending more than 1 cm above the sporangium, sporangia without colour. - *Isoetes bolanderi*
- the end segment of the branchlets is 1-celled.

Section 2: Flowering Plants (Angiosperms)

In *Nuphar* the large, showy, yellow, perianth members are sepals, the petals are smaller than the stamens and inconspicuous.

Mature fruits are generally needed for positive identification of *Callitriche.*

- leaves peltate, petiole attached inside the margin, and floating on the surface.
- alternate leaves arise near the apex of a vertical floating stem, leaves and plant covered in a gelatinous sheath, leaves entire, oval and reddish, flowers solitary, reddish, on long peduncles from the upper axils, deep water plant, fruit a follicle. - *Brasenia schreberi*
- leaves solitary from the nodes of a horizontal creeping rhizome, no gelatinous sheath, leaves rounded, margins not entire, leaves green, flowers in umbels on a long peduncles from the nodes, shallow water plants, fruit a schizocarp. - *Hydrocotyle verticillata*
- leaves not peltate, petiole attached on the leaf margin, there may be a deep sinus with leaves apparently peltate in some Nymphaea.
- leaves arise in basal clusters, (plants scapose) or from relatively short, usually apical, sections of rhizomes, but not cauline and spread along the stem.
- leaves all fully submersed and much longer than wide.
- leaves with long petioles and lanceolate blades, solitary flowers on stalks shorter than the leaves, arising in a cluster among the leaves. - *Limosella aquatica*
- leaves sessile, flowers on stalks, exceeding or equalling the leaves.
- leaves very long and tape-like, flowers solitary on very long peduncles which reach the surface and coil up like a spring after fertilization and draw the fruit underwater. - *Vallisneria americana*
- leaves shorter than the flower stalks, flowers on few-flowered racemes, no coiled peduncle.

- leaves linear and sub-terete, not fleshy, several few flowered racemes, roots not white and conspicuous. - *Subularia aquatica*
- leaves thick and fleshy, terete or angular, 1 few-flowered raceme, plants with a dense cluster of white roots. - *Lobelia dortmanna*
- leaves, or some at least, with long petioles and floating on the surface, blades nearly round and with a deep sinus.
- tubers or slender rhizomes, 5 petals and sepals united at the base, leaves small, under 15 cm in diameter, flowers white or yellow. - *Nymphoides*
- flowers yellow, axillary, no cluster of roots on the petiole, leaves primarily from branching stems. - *Nymphoides peltata*
- flowers white, in clusters on the petioles, a cluster of roots on the petiole, leaves mostly basal with long slender petioles.
- leaf blades ovate, mostly under 5 cm long, seeds smooth, leaves not dark-punctate beneath. - *Nymphoides cordatum*
- leaf blades orbicular, 8 to 15 cm in diameter, seeds glandular, warty, leaves dark-punctate or pitted beneath. - *Nymphoides aquaticum*
- stout or massive rhizomes, numerous petals, 5 to 12 sepals, leaves large, some over 15 cm in diameter, flowers white, yellow, red, or pink.
- flowers yellow, 5 to 12 sepals, outer ones green, inner ones yellow tinged with red or green, no arils on the seeds, superior ovary, petals erect and smaller than the sepals or stamens, 1 discoid stigma. - *Nuphar*
- Six to 8 (usually 6) sepals, 2.5 to 3.5 cm long, yellow stamens. - *Nuphar variegatum*
- Eight to 17 (usually 9) sepals, (3) 3.5 to 6 cm long, reddish stamens. - *Nuphar polysepalum*
- flowers white, red or (yellow), sepals 4, green or streaked with red, seeds arillate, semi-inferior ovary, petals spreading, many stigmas. - *Nymphaea*
- flowers yellow, 6 to 13 cm in diameter, leaves dark green and blotched above, brownish with black dots below, plant tuberous. - *Nymphaea mexicana*
- flowers white pink or red, leaves not blotched.
- small plant with leaves up to 8 cm wide and flowers up to 5 cm wide, 7 to 15 petals, 6 to 9 stigmas, flowers white.

- carpellary appendages 3mm or longer, flaccid and purplish. - *Nymphaea tetragona*
- carpellary appendages 1.5mm or shorter, stiff and greenish. - *Nymphaea leibergii*
- larger plants with leaves over 8 cm wide and flowers over 5 cm wide, petals and stigmas numerous, flowers white, pink, or rose.
- flowers white or tinged with pink, 20 to 32 petals, leaves scattered on the rhizome, flowers strongly scented, leaves usually purplish beneath. - *Nymphaea odorata*
- flowers white, 12 to 24 petals, leaves crowded on the rhizome, flowers not strongly scented, leaves greenish beneath. - *Nymphaea alba*
- leaves cauline and more or less evenly distributed along the stem.
- horizontal, creeping rhizomes or stolons with leaves arising at the nodes, often rooted at every node.
- leaves bladeless and sessile, clustered, terete, septate, glabrous, flowers in short stalked umbels, shorter than the leaves and arising among the leaves at the nodes, stolons creeping. - *Lilaeopsis occidentalis*
- leaves with a blade and a petiole.
- solitary deeply lobed circular leaves at each node, small inconspicuous flowers in short-stalked, recurved umbels, one umbel at each node, stolons tending to creep. - *Hydrocotyle ranunculoides*
- leaves lobed and solitary to lanceolate and several, flowers solitary in the leaf axils of upper nodes, conspicuous and yellow or white, stolons creeping or arched between nodes. - Ranunculaceae
- several lanceolate, entire leaves at each node, stolons often arched between nodes, (apical leaves may be sessile), flower stalks longer than the apical leaves but shorter than the long-petiolate basal leaves. - *Ranunculus flammula*
- leaves solitary and deeply lobed at most nodes, stolons mostly flat, flower stalks about as long as the leaves.
- flowers yellow, sepals and petals, leaf margins entire, leaves 3 -lobed, 1 to 1.5 cm wide, fruit an ovoid head of achenes. - *Ranunculus hyperboreus*

- flowers white, sepals but no petals, leaf margins dentate, leaves reniform, 2 to 4 cm wide, fruit a cluster of follicles. - *Caltha natans*
- more or less erect or sprawling stems, not buried or rooted at every node.
- leaves opposite or whorled, several per node (in juvenile *Najas* with short internodes this may be difficult to determine).
- leaves opposite, 2 at a node.
- leaves capillary with membranous, sheathing stipules, stigma peltate, fruits clusters of short-stalked, banana-shaped, conspicuous achenes in the leaf axils, 1 stamen with a filament longer than the pistils, stems filiform and freely branched. - *Zannichellia palustris*
- leaves at least linear and sometimes ovate, no stipules, stigmas not peltate, fruit not a cluster of banana-shaped achenes, stamens various, stems not filiform and branched or unbranched.
- leaf with a clasping, short, broad base abruptly narrowing to long-attenuate, toothed on the shoulder, fruits solitary, fusiform, smooth and shining, achenes in the leaf axils, stamen almost sessile, plants branched and bushy apically and fragile, often fragmenting. - *Najas flexilis*
- leaves linear to obovate, not with a sheathing base and not abruptly narrowing, fruits axillary and bi-lobed or several-seeded capsules, stamens solitary with a long filament or 9 on stalked flowers, stems often unbranched or with a few apical branches, not usually bushy.
- leaves linear to narrowly lanceolate, 1-nerved, usually paired basally and whorled apically, 9 stamens, long-stalked flowers, fruit of several-seeded capsules. - *Elodea*
- leaves rarely over 1.5 cm long.
- leaves 2 (1 to 4) mm wide, tapered abruptly to a blunt point, staminate flowers stalked and persistent. - *Elodea canadensis*
- leaves 1.5 (0.3 to 1.5) mm wide, tapered to a slender point, staminate flowers sessile and deciduous at anthesis. - *Elodea nuttallii*
- leaves (1.7) 2.0 to 2.6 cm long. - {*Elodea longivaginata*}
- leaves linear to obovate or spathulate, 1 to 3 (5) nerved, all opposite, sometimes linear basally and obovate and clustered apically, fruit of bi-lobed, often winged, sessile, axillary, naked achenes. - *Callitriche*

- fruit encircled by a conspicuous wing-like margin, leaf bases joined by winged ridges. - *Callitriche stagnalis*
- fruit not winged or with only a narrow wing at the tip, leaf bases various.
- leaves all linear, 1-nerved, light green, leaf bases not joined by a wing, floral bracts absent, common species. - *Callitriche hermaphroditica*
- leaves various, upper often ovate and 3-nerved, bases joined by a wing-like ridge, floral bracts present.
- carpel face markings in regular vertical lines, fruit slightly wing-margined at the top and longer than broad. - *Callitriche verna*
- carpel face markings scattered, fruit not winged and as long as broad.
- fruits widest above the middle (obovate) leaves bidentate, midvein barely thickened at the end, emergent leaves may be over 5 mm wide, stems long. - *Callitriche heterophylla*
- fruits round or oblong, midvein thickened at the, tip and protruding, leaves never over 5 mm wide, plants short and slender. - *Callitriche anceps*
- leaves whorled, 3 or more at a node.
- leaves (8) 10 to 25 (50) mm long in whorls of (4) 6 to 12, 1 style, 1 stamen, fruit nut-like, flowers solitary and sessile in the axils, glabrous perennials from creeping rhizomes, the apical portion of the stem may be emergent or the plant may grow completely emergent on wet mud with much shorter leaves and shorter internodes. - *Hippuris vulgaris*
- leaves under 25 (40) mm long in whorls of (2) 3 to 12, 3 or more styles, 3, 6 or 9 stamens, fruit a few-seeded capsule, flowers axillary and sessile but on long, thin stalks, the hypanthium, which reaches the surface, permanently submersed. - Hydrocharitaceae
- leaf arrangement irregular, margins serrate or toothed, tip with 2 enlarged spines. - {*Lagarosiphon major*}
- leaves arranged in regular whorls.
- leaf margins serrate or toothed.
- leaf tip with 2 enlarged spines, no turions , 3 or more strongly recurved leaves up to 3 cm long per whorl. - {*Lagarosiphon major*}

- no enlarged spines on the leaf tip, turions present on the rhizome or on the stem, 3 to 8 (12) nearly straight leaves up to 2.5 cm long per whorl. - {*Hydrilla verticillata*}
- leaf margins entire.
- basal leaves in whorls of 3 but upper ones up to 6 per whorl and up to 4 cm long, 2 to 3 flowers in the staminate spathes, petals to 10 mm. - {*Egeria densa*}
- basal leaves in pairs, the upper ones usually in whorls of 3 and up to 2.5 cm long, flowers solitary in the staminate spathes, petals up to 5 mm long. - *Elodea*
- leaves in whorls of 3.
- leaves rarely over 1.5 cm long.
- leaves 2 (1 to 4) mm wide, tapered abruptly to a blunt point, staminate flowers stalked and persistent. - *Elodea canadensis*
- leaves 1.5 (0.3 to 1.5) mm wide, tapered to a slender point, staminate flowers sessile and deciduous at anthesis. - *Elodea nuttallii*
- leaves (1.7) 2.0 to 2.6 cm long. - {*Elodea longivaginata*}
- leaves in the upper and middle part of the stem in pairs, lower ones irregular or alternate, (1.7) 2.0 to 2.6 cm long. - (*Elodea longivaginata*)
- plants sprawling at or just under the surface, nodally rooting, flowers showy, rose in terminal spicate panicles or pale yellow and solitary from the upper axils.
- flowers rose in a compact, terminal panicle, leaves broadly lanceolate to elliptic, floating on the surface, petiolate, leathery. - *Polygonum amphibium*
- flowers pale yellow, solitary on long stalks, in the axils of apical leaves, submersed, long and narrow, delicate. - *Heteranthera dubia*
- plants mostly erect, not usually rooting at the nodes, flowers reduced and in whorls on pedunculate spikes, inconspicuous. - *Potamogeton*
- stipules forming a sheath around the stem partly below the base of the leaf blade and partly above, the leaf attached 1 cm or more above the node.
- leaves 3 to 4 mm wide, flat, rough on the edges, stiff, with a broad mid-vein and over 20 lateral veins. - *Potamogeton robbinsii*

- leaves less than 3 mm wide, cylindrical, smooth, soft, with only one inconspicuous vein.
- leaves acute and sharp pointed at the apex, fruits 2.5 to 4 mm long with a short beak, the surface of fruits without radial striation (at 10x). - *Potamogeton pectinatus*
- leaves obtuse and blunt at the apex, fruits 2 to 3 mm long, beakless, surface of the fruits with radial striation (at 10x).
- primary stems 0.5 to 1 mm in diameter, all leaves filiform, 0.2 to 0.5 mm wide with tight sheaths, spikes with 2 to 5 whorls of flowers, fruits 2 to 2.5 mm long. - *Potamogeton filiformis*
- primary stems 1 to 3 mm in diameter, lower main stem leaves with blades 1 to 2 mm wide and with loose inflated sheaths, upper branch leaves filiform, spikes with 5 to 12 whorls of flowers, fruits 2.5 to 2.8 mm long. - *Potamogeton vaginatus*
- stipules forming a sheath around the stem only above the base of the leaf blade, the leaf attached at the node.
- submersed leaves linear, ribbon-like or cylindrical, less than 5 mm wide, parallel margins.
- submersed leaves cylindrical, terete.
- floating leaves 2.5 to 6 cm wide, usually cordate, large ones with (18) 21 to 35 nerves, submersed leaves arise from the main stem, mature fruit 3.7 to 4.5 mm long including the beak, obscurely keeled, interlacunar bundles in several circles throughout the aerenchyma. - *Potamogeton natans*
- floating leaves 1 to 3 cm wide, usually rounded or cuneate at the base, largest ones with 11 to 19 nerves, submersed leaves arise from the branches of the main stem, mature fruit, including the beak, 3 to 3.4 mm long, interlacunar bundles develop only in the outer circle of the aerenchyma. - *Potamogeton oakesianus*
- submersed leaves flat.
- leaves alternate, 1 per node.
- submersed leaves with 9 to 35 veins. - *Potamogeton zosteriformis*
- submersed leaves with 1 to 7 veins.
- stipules strongly fibrous, becoming whitish, especially on the turions, base of turions strongly ribbed.
- leaves obtuse or rounded and slightly mucronate, not conspicuously 2 ranked,, blades thin, 1.5 to 3.5 mm wide, 5 to

7 veins, a narrow cellular-reticulate band along the midrib, turions fan-shaped, peduncle flattened. - *Potamogeton friesii*

- leaves gradually tapered into a sharp bristle tip, conspicuously 2 ranked, blades firm, 0.5 to 2.5 mm wide, convolute with 3 (5) veins, no cellular-reticulate band along the midrib, turions slender, peduncles terete. - *Potamogeton strictifolius*
- stipules delicate, not fibrous, greenish or brownish, base of the turions smooth.
- leaves (2) 3 to 4 mm wide, rounded at the apex, fruits (3) 3.5 to 4 mm long. - *Potamogeton obtusifolius*
- leaves 0.3 to 3 mm wide, acute to obtuse or mucronate, fruits 1.8 to 2.8 mm long
- fruits with a distinct dorsal keel, veins on the stipules evident as ridges running the full length of the stipule, glands at the base of the stipules either lacking or poorly developed. - *Potamogeton foliosus*
- fruits without a dorsal keel, veins on the stipules obscure and faint, glands at the base of the stipules usually well developed.
- stipules connate, fused along the stem, at least when young, mature fruits broadest above the middle, plants sparsely branched. - *Potamogeton pusillus*
- stipules convolute, mature fruits broadest below the middle, plants well branched. - *Potamogeton berchtoldii*
- submersed leaves lanceolate or ovate, over 5 mm wide.
- leaf margins serrate, beak of the fruit as long as the fruit body or longer. - *Potamogeton crispus*
- leaf margins entire, beak of the fruit much shorter than the fruit body.
- submersed leaves ribbon-like with parallel side, 5 to 10 mm wide, limp and flaccid, median band of lacunae several cells wide at least 1/4 the width of the blade, stems compressed. - *Potamogeton epihydrus*
- submersed leaves lanceolate to ovate with sides not parallel, either no median cellular-reticulate band or the band less than 1/4 the width of the blade, stems terete.
- submersed leaves sessile, cordate or rounded at the base and clasping the stem.

- leaves ovate-oblong, mostly 10 to 20 cm long, with a cucullate apex, fruits more than 4 mm long, stems with many interlacunar vascular bundles. - *Potamogeton praelongus*
- leaves rounded, ovate or elongate-ovate, 1 to 10 cm long, no cucullate apex, fruits under 3.5 mm long, no interlacunar vascular bundles.
- stipules coarse disintegrating into persistent whitish fibres, peduncles clavate, 1.5 to 25 cm long, fruits with a cavity in the endocarp loop. - *Potamogeton richardsonii*
- stipules delicate, lacking on mature specimens, peduncles not clavate, 1 to 9 cm long, fruits without a cavity in the endocarp loop. - *Potamogeton perfoliatus*
- submersed leaves petiolate or sessile, not clasping the stems.
- submersed leaves 2 to 5 cm wide, tapering into distinct petioles, stipules over 3 cm long.
- floating leaves cordate, submersed leaves folded and strongly falcate, more than 20 veins, interlacunar bundles well developed throughout. - *Potamogeton amplifolius*
- floating leaves rounded or cuneate at the base, submersed leaves flat and not falcate, with fewer than 20 veins, interlacunar bundles developed in only one circle or absent.
- petioles of submersed leaves shorter than 2 cm, interlacunar bundles well developed forming one circle, endodermis of U-cells. - *Potamogeton illinoensis*
- petioles of submersed leaves longer than 2 cm, interlacunar bundles absent, endodermis of O-cells. - *Potamogeton nodosus*
- submersed leaves mostly less than 2 cm wide, sessile, or with petioles under 0.5 cm long, stipules under 3 cm long.
- plants with a reddish tinge, usually not branched, floating leaves, if present, not markedly different from the submersed leaves, no interlacunar bundles, endodermis of O-cells. - *Potamogeton alpinus*
- plants greenish, freely branched, floating leaves markedly different from the submersed leaves, interlacunar bundles well developed in one circle, endodermis of U-cells. - *Potamogeton gramineus*

Part 4: Emergents- Opposite or Whorled Cauline Leaves

Our aquatic *Equisetum* all share the following characteristics: stems annual, usually with whorls of branches, distinct sterile and fertile stems, cones blunt.

- leaves in whorls of 3 or more at a node, small and scale or bract-like, basally sheathing, erect stems with longitudinal grooves offset at each node, sporangia in terminal cones, non-flowering plants, horsetails. - *Equisetum*
- central cavity well over 1/2 the diameter of the stem, teeth may be deciduous, 10 to 40 ridges, teeth persistent, black but not hyaline margined, stomates in 1 row and not sunken. - *Equisetum fluviatile*
- central cavity less than 1/3 the diameter of the stem, teeth not deciduous, black and hyaline margined, 5 to 10 ridges. - *Equisetum palustre*
- leaves opposite, only 2 at a node, flowering plants.
- flowers sessile in the axils, 2 to 4-merous perianth, capsules.
- leaves under 1.5 cm long, spathulate, flowers (2) 3 merous, seeds curved and pitted. - *Elatine rubella*
- leaves 2 to 6 cm long, ovate to elliptic, no petals, flowers 4-merous, seeds glabrous. - *Ludwigia palustris*
- flowers on stalks in the axils.
- plants sprawling or creeping, if more or less erect under 1 dm tall, not woody or spongy at the base, flowers not in whorls in the axes or in congested racemes (*Lysimachia thyrsiflora*).
- flowers solitary in the axils.
- five sepals, 2-lobed corolla tube, 2 stamens and stigmas, fruit a capsule. - *Gratiola*
- there are 2 sepaloid bracts at the apex of the pedicel and thus apparently 7 sepals. - *Gratiola neglecta*
- pedicels without bracts and sepals thus evidently only 5. - *Gratiola ebracteata*
- flowers greenish, 4 sepals and petals, 4 stamens, fruit of 4 follicles. - *Crassula aquatica*
- flowers in open racemes, 2 sepals, 3 stamens, 3 stigmas, sprawling and creeping, rooting at the nodes, annual. - *Montia fontana*

- plants more or less erect to apically drooping, over 1 dm tall.
- flowers crowded in short, dense racemes axillary in the middle leaves, flowers 5 (6 to 7) merous. - *Lysimachia thyrsiflora*
- flowers in elongate, open racemes, 4 (5) sepals, corolla 4 lobed, 2 stamens, capitate stigma. - *Veronica*
- leaves all short-petiolate. - *Veronica beccabunga*
- leaves, at least the middle and upper ones, sessile and clasping.
- capsules much wider than high, notched at the tip,, leaves (three) 4 to 20 times as long as wide. - *Veronica scutellata*
- capsules not wider than high, scarcely, if at all notched at the tip,, leaves 1.5 to 5 times as long as wide.
- leaves 1.5 to 3 times as long as wide, fruiting pedicels ascending or upcurved, flowers blue or violet. - *Veronica* anagallis-aquatica
- leaves (2.5) 3 to 5 times as long as wide, fruiting pedicels spreading, flowers white to pink or pale blue. - *Veronica catenata*

Part 5: Emergents- Alternate Cauline Leaves

- leaves compound.
- flowers in umbels, (Apiaceae or Umbelliferae, base of the stem without transverse septa, roots not tuberous thickened, ribs of the fruit prominent, corky-thickened or covered in prickles, primary lateral veins of the leaflets not directed in any particular direction, plants erect, calyx teeth tiny or absent). - *Sium sauve*
- flowers in axillary racemes, (fruit a silique, plants prostrate and nodally rooting, plant used as a salad green, watercress, seeds biseriate). - *Nasturtium officinale*
- leaves simple.
- leaf margins lobed or toothed.
- leaves lobed, flowers yellow, fruit a head of achenes, inflorescence axillary. - *Ranunculus hyperboreus*
- leaves only toothed or crenulate, long-petiolate, rounded with a deep basal sinus, margin crenate, flowers solitary on long peduncles. - *Caltha*
- sepals yellow, no petals, plants erect or sprawling. - *Caltha palustris*
- sepals white, no petals, plants creeping or floating. - *Caltha natans*

- leaf margins entire or nearly so.
- flowers naked or subtended by chaffy or bristle-like bracts.
- flowers mostly perfect, if imperfect then in reduced spikes with the staminate often not uppermost, grasses or sedges
- leaves in 3 vertical ranks on an often triangular, solid or pithy, stem without swollen nodes, sheaths closed, flowers subtended by 1 (2) bract (s) and often with inner subtending scales or bristles, fruit a usually beaked achene, ovary sometimes enclosed in a sac-like covering, styles often 3. - Cyperaceae
- leaves in 2 vertical ranks on a usually terete, hollow stem with swollen nodes, sheaths commonly open but may be closed, flowers usually subtended by 2 bracts with 2 obscure inner subtending scales, ovary never enclosed in a sac, fruit a grain with the ovary wall adherent to the seed, rarely beaked, up to 2 styles, grasses. - Poaceae
- flowers mostly imperfect, several in globose to capitate clusters, each subtended by 3 to 5 chaffy bracts, fruit hard, beaked, conspicuous, staminate and pistillate in separate heads not grass-like (bur-reeds). - *Sparganium*
- stigmas over 2 mm long, the achenes truncate-pyriform and narrowed abruptly to the beak, inflorescence usually branched, staminate heads above the pistillate heads, large robust plants. - *Sparganium eurycarpum*
- stigma, achenes fusiform and tapering to the beak, inflorescence usually simple, staminate heads below the pistillate heads.
- achene beaks less than 1.5 mm long or beakless, staminate head solitary.
- achene beaks less than 0.5 mm long to absent, staminate head closely adjacent to the upper pistillate head, basal leaves (0.5) 1 to 3 (5) mm wide, opaque, often yellow. - *Sparganium hyperboreum*
- achene beaks 0.5 to 1.5 mm long, staminate head distant from the upper pistillate head, basal leaves (1.5) 2 to 6 (10) mm wide, translucent, usually dark green. - *Sparganium natans*
- achene beaks 1.5 to 5 mm long, there may be more than 1 staminate head.
- achene beaks conspicuously curved. - *Sparganium fluctuans*
- leaves parallel veined, floral parts in 3's (4's) but not 5's. - Monocotyledons

- achene beaks straight or slightly curved.
- staminate head usually solitary, closely adjacent to the upper pistillate head. - *Sparganium glomeratum*
- staminate heads (1) 3 to 8, distant from the upper pistillate head.
- pistillate heads (1) 2 to 3 (4), usually crowded and appearing as one elongated head terminating the inflorescence, achene beaks 1.5 to 2.0 (2.2) mm long. - *Sparganium angustifolium*
- pistillate heads (3) 4 to 7 (10), distant and distinct, achene beaks 2 to 4.5 (6) mm long. - *Sparganium emersum*
- flowers with a perianth, all alike or with distinct sepals and petals.
- flowers imperfect, several in globose to capitate clusters, each subtended by 3 to 5 chaffy bracts, fruit hard, beaked, conspicuous, (bur-reeds). - *Sparganium*
- flowers perfect or if imperfect (rare) then staminate and pistillate flowers not in separate capitate clusters and not subtended by 3 to 5 chaffy bracts, fruit various.
- inflorescence a terminal fleshy spike subtended by a large, usually whitish, showy spathe, flowers closely crowded and embedded, perianth and stamens in 4's, large red berries. - *Calla palustris*
- inflorescence not a fleshy spike with embedded flowers and a spathe, perianth and stamens usually other than 4.
- leaves in 3 vertical ranks on an often triangular, solid or pithy, stem without swollen nodes, sheaths closed, flowers subtended by 1 (2) bract (s) and often with inner subtending scales or bristles, fruit a usually beaked achene, ovary sometimes enclosed in a sac-like covering, styles often 3. - Cyperaceae
- leaves mostly not 3-ranked, stems usually terete, rarely grass-like, perianth segments usually in 2 series of 3 each.
- pistils more than 1, distinct or separating at maturity into 3 to 6 follicles, perianth inconspicuous, greenish, segments all alike, 3 to 6 carpels.
- flowers in a 3 to 12 flowered bracteate raceme, fruiting pedicels 12 to 25 mm long, follicles compressed, divergent, 1 or 2 seeded. - *Scheuchzeria palustris*
- flowers in more than 12 flowered, bracteate racemes or spikes, fruiting pedicels rarely up to 6 mm long, follicles neither

compressed nor divergent, 1 seeded, no perianth, pistillate flowers axillary, perfect flowers in pedunculate spikes, long styles, fruit 3-lobed, an annual plant. - *Lilaea scilloides*

- 1 compound pistil, 1 or 3 celled, fruit not a follicle.
- plants grass-like, flowers inconspicuous, perianth greenish, purplish or brownish, segments all alike. - *Juncus supiniformis*
- plants not grass-like, flowers showy, blue or purple, 6 stamens, superior ovary, long-petiolate, sheathing, arrow-shaped, hastate, leaves, perianth fused, inflorescence subtended by 2 spathes, introduced. - *Pontederia cordata*
- leaves pinnately or palmately veined, flower parts mostly in 5's, sometimes 4's, rarely 3's. - Dicotyledons
- leaves long-petiolate, often longer than the blade which is lanceolate, plants creeping and nodally rooting, fruit a globose cluster of achenes, flowers yellow. - *Ranunculus flammula*
- leaves sessile or short-petiolate, plants usually more or less erect but if creeping or nodally rooting then the flowers blue in a lax, helicoid, raceme.
- leaves with generally lacerate, sheathing stipules, nodes swollen, 2 or 3 styles, (4 or 6) 8 stamens. - *Polygonum* (in part)
- perianths glandular-punctate, glands sessile, achenes brown, glandular and dull, perianth usually 4-lobed, stamens usually 4 or 6. - *Polygonum hydropiper*
- perianths not glandular-punctate, achenes dark brown or black, perianth 4 to 5 lobed, stamens 8.
- inflorescence ending in spicate, slender, interrupted racemes mostly over 3 cm long, nerves of the perianth segments not branched and recurved, stamens 8, perianth 5-lobed. - *Polygonum hydropiperoides*
- inflorescence of several short, thick, continuous racemes, rarely over 3 cm long, nerves of the perianth segments branched and recurved, perianth 4 to 5 lobed. - *Polygonum lapathifolium*
- leaves without lacerate, sheathing stipules, nodes not swollen, one style, 5 stamens, flowers in helicoid, sympodial false racemes (forget me not). - *Myosotis*
- corolla limb 2 to 5 mm wide, style shorter than the nutlets, not stoloniferous, often recumbent. - *Myosotis laxa*
- corolla limb 5 to 10 mm wide, style as long as or longer than the nutlets, stoloniferous, mostly erect and not creeping or recumbent. - *Myosotis scorpioides*

Part 6: Emergents- Basal or Terminal Leaf Clusters

In *Sagittaria* the usual non-achene key characters of bract length and shape and pedicel lengths have been found to be inconsistent and sufficiently variable even within one population as to be virtually useless as key characters. Many flowering, but not fruiting, collections are difficult to identify.

- leaves compound, flowering plants, with several leaves, leaves trifoliate, leaflets obovate and entire to undulate-dentate, petals white and densely scaly or hairy, flowers in racemes, fruit a capsule. - *Menyanthes trifoliata*
- leaves simple.
- non-flowering plants, spores, no flowers, quillworts, cluster of basally sheathing, narrow, terete leaves with spores in the axils of the expanded base. - *Isoetes nuttallii*
- flowering plants, with several leaves, flowers and seeds are formed, not spores, leaves not as above or with sporangia in the axils.
- leaves parallel veined, floral parts in 3's (4's) but not 5's, (*Calla* and *Pontederia* do not have parallel-veined leaves but have spathes). - Monocotyledons
- flowers naked or subtended by chaffy or bristle-like bracts.
- flowers mostly perfect, if imperfect then in reduced spikes with the staminate often not uppermost, grasses or sedges
- leaves in 3 vertical ranks on an often triangular, solid or pithy, stem without swollen nodes, sheaths closed, flowers subtended by 1 (2) bract(s) and often with inner subtending scales or bristles, fruit a usually beaked achene, ovary sometimes enclosed in a sac-like covering, styles often 3. - Cyperaceae
- leaves in 2 vertical ranks on a usually terete, hollow stem with swollen nodes, sheaths commonly open but may be closed, flowers usually subtended by 2 bracts with 2 obscure inner subtending scales, ovary never enclosed in a sac, fruit a grain with the ovary wall adherent to the seed, rarely beaked, up to 2 styles, grasses. - Poaceae
- flowers mostly imperfect, several in globose to capitate clusters, each subtended by three to 5 chaffy bracts, fruit hard, beaked, conspicuous, staminate and pistillate in separate heads, not grass-like, (bur-reeds). - *Sparganium*

- two (one) stigmas over 2 mm long, the achenes truncate-pyriform and narrowed abruptly to the beak, inflorescence usually branched, staminate heads above the pistillate heads, large robust plants. - *Sparganium eurycarpum*
- one stigma, achenes fusiform and tapering to the beak, inflorescence usually simple, staminate heads below the pistillate heads.
- achene beaks less than 1.5 mm long or beakless, staminate head solitary.
- achene beaks less than 0.5 mm long to absent, staminate head closely adjacent to the upper pistillate head, basal leaves (0.5) 1 to 3 (5) mm wide, opaque, often yellow. - *Sparganium hyperboreum*
- achene beaks 0.5 to 1.5 mm long, staminate head distant from the upper pistillate head, basal leaves (1.5) 2 to 6 (10) mm wide, translucent, usually dark green. - *Sparganium natans*
- achene beaks 1.5 to 5 mm long, there may be more than one staminate head.
- achene beaks conspicuously curved. - *Sparganium fluctuans*
- achene beaks straight or slightly curved.
- staminate head usually solitary, closely adjacent to the upper pistillate head. - *Sparganium glomeratum*
- staminate heads (1) 3 to 8, distant from the upper pistillate head.
- pistillate heads (1) 2 to 3 (4), usually crowded and appearing as one elongated head terminating the inflorescence, achene beaks 1.5 to 2.0 (2.2) mm long. - *Sparganium angustifolium*
- pistillate heads (3) 4 to 7 (10), distant and distinct, achene beaks 2 to 4.5 (6) mm long. - *Sparganium emersum*
- flowers with a perianth, all alike or with distinct sepals and petals.
- flowers imperfect, several in globose to capitate clusters, each subtended by 3 to 5 chaffy bracts, fruit hard, beaked, conspicuous, (bur-reeds). - *Sparganium*
- flowers perfect or if imperfect (rare) then staminate and pistillate flowers not in separate capitate clusters and not subtended by 3 to 5 chaffy bracts, fruit various.
- inflorescence a terminal fleshy spike subtended by a large, usually whitish, showy spathe, flowers closely crowded and

embedded, perianth and stamens in 4's, large red berries. - *Calla palustris*

- inflorescence not a fleshy spike with embedded flowers and a spathe, perianth and stamens usually other than 4.
- leaves in 3 vertical ranks on an often triangular, solid or pithy, stem without swollen nodes, sheaths closed, flowers subtended by one (2) bract (s) and often with inner subtending scales or bristles, fruit a usually beaked achene, ovary sometimes enclosed in a sac-like covering, styles often 3. - Cyperaceae
- leaves mostly not 3-ranked, stems usually terete, rarely grass-like, perianth segments usually in 2 series of three each.
- pistils more than 1, distinct or separating at maturity into 3 to six follicles
- perianth inconspicuous, greenish, segments all alike, 3 to 6 carpels.
- flowers in a 3 to 12-flowered bracteate raceme, fruiting pedicels 12 to 25 mm long, follicles compressed, divergent, 1 or 2 seeded. - *Scheuchzeria palustris*
- flowers in more than 12-flowered, bracteate spikes, fruiting pedicels rarely up to 6 mm long, follicles neither compressed nor divergent, 1 seeded, no perianth, pistillate flowers axillary, perfect flowers in pedunculate spikes, long styles, fruit 3-lobed. - *Lilaea scilloides*
- perianth showy, sepals green, petals white to pinkish, 10 or more carpals, fruit an achene, stamens 6 (9) or over 15, inflorescence a bracteate panicle or raceme. - Alismataceae
- leaves basally sagittate or hastate, 7 to 25 stamens, achenes densely packed on the receptacle. - *Sagittaria*
- mature achenes 2.0 to 2.5 mm long with a beak less than 0.5 mm long pointing forward from the tip of the achene, monoecious. - *Sagittaria cuneata*
- mature achenes (2.5) 3.0 to 3.5 (4.0) mm long with a beak about 1 mm long at right angles to the body of the achene, monoecious or dioecious. - *Sagittaria latifolia*
- leaves gradually tapered or cordate at the base; broadly ribbon-like if submersed.
- leaves completely submersed in deep water or floating on the surface, ribbon-like or tapered at the base and oval, flaccid, plants generally sterile. - *Sagittaria*

- leaves emergent and erect, ovate to lanceolate, stamens 6 (9), achenes in a single whorl on the receptacle. - *Alisma*
- leaf blades lanceolate to ovate, scapes longer than the leaves (petiole and blade), achenes centrally grooved apically. - *Alisma* plantago-aquatica
- leaf blades narrowly lanceolate to linear, scapes shorter than the leaves (petiole plus blade), achenes 2-grooved at the tip. - *Alisma gramineum*
- compound pistil, 1 or 3 celled, fruit not a follicle.
- plants grass-like, flowers inconspicuous, perianth greenish, purplish or brownish, segments all alike. - *Juncus supiniformis*
- plants not grass-like, flowers showy, blue or purple, 6 stamens, long-petiolate, arrow-shaped, hastate leaves, perianth fused, inflorescence subtended by 2 spathes, introduction. - *Pontederia cordata*
- leaves not long and grass-like with parallel veins, flower parts not in 3's, usually in 5's, flowers yellow. - Dicotyledons (Ranunculaceae)
- leaves rounded to reniform with crenate margins, sepals but no petals, stem becoming creeping and nodally rooting, flowers solitary on long stalks. - *Caltha palustris*
- leaves rhombic to chordate-rotund, all in a basal cluster, plants stoloniferous, petals with a basal pocket or nectary scale, flowers few to solitary on long stalks. - *Ranunculus cymbalaria*

Part 7: Marine Plants

- plants generally subtidal (may be briefly exposed at the lowest tides), long, narrow, grass-like leaves, many flowers embedded in a fleshy spadix, no elongate, coiled peduncle when fruiting. - Zosteraceae
- leaf sheaths deciduous, sometimes leaving a few scaly parts behind, leaf blades thin and translucent, rhizome with elongate internodes (1 to 3 cm long or more), 2 thin roots at each internode, monoecious, spadix border projections inconspicuous, usually established on sand or mud. - *Zostera*
- leaves 3-veined, 1 to 1.5 mm wide, sheaths split to the base, rare and probably introduced from Asia, known only from Boundary Bay and Tsawwassen. - *Zostera japonica*
- leaves 5-veined or more, 1.5 to 12 mm wide, sheaths on sterile shoots are closed at the base, widespread native species. - *Zostera marina*

- basal portions of the leaf sheaths decay with age to bundles of fine, woolly fibres, leaf blades leathery, rhizome has short, thick internodes with 2 or more thick roots at each internode, dioecious, spadix bordered by conspicuous flap-like projections, usually established rocks. - Phyllospadix
- fertile stems branched, 4 to 12 dm long, spathes usually paired at the nodes. - *Phyllospadix torreyi*
- fertile stems unbranched, 0.5 to 4 dm long, spathes usually solitary at the nodes.
- leaves with 3, rarely 5, veins, margins entire. - *Phyllospadix scouleri*
- leaves with 5 or 7 veins, margins toothed towards the apex. - *Phyllospadix serrulatus*
- plants submersed in brackish ditches and lagoons, long, narrow, grass-like leaves, two naked flowers on a short spadix arising from a sheath of leaf bases, 2 stamens and 4 pistils which become a cluster of fruits, peduncle and stipes elongate and coil at maturity, submersed. - *Ruppia maritima*

Classification of Aquatics by Life-form and Habitat

From an ecological point of view the growth-form of a plant and its usual habitat are often more important than the specific identification. Communities can often be distinguished by the growth-form of the plants present, which is constant world-wide, while the species may vary from place to place. Due to the morphological plasticity of aquatic plants no classification can be more than approximate and many exceptions will be found. A number of such classification schemes exist for various purposes. The following one may prove useful in reducing the number of choices when trying to identify an unknown plant, and in defining which group of plants normally inhabit certain zones. The key has been written to determine which group of plants is present, and under each group code mentioned in the key there is a list of species or genera which are applicable. Some plants will appear in more than one group. Generally only intact mature plants are keyed; fragments and seedlings would cause confusion.

Key to Mature Aquatic Plants of British Columbia: Based on Growth-forms and Habitat

- Floating freely in or on the water; attached only when stranded by lower water levels.
- Floating on the surface of the water.

- Large plants with distinct roots, stems and leaves; may have an obvious swollen part used as a floatation organ. - Group A
- Small thalloid plants with roots below; closely appressed to the surface of the water. - Group B
- Floating beneath the surface of the water.
- Found just under the surface tension of the water or free in the mid-water zone; often found tangled in other submersed or marginal vegetation. - Group C
- Found lying on or just above the sediment surface. - Group D
- Rooted or anchored in the sediment at all water levels. (fragments of *Elodea*, {*Egeria densa*}, *Najas flexilis, Ranunculus, Callitriche, Ceratophyllum*, {*Limnophila sessiliflora*} and *Myriophyllum* may be found adrift with Group C).
- All vegetative parts remain beneath the surface of the water. (several species begin the season submersed but as water levels drop end up in Group O; these are primarily plants of Group E and Group G but include a few members of Group I and Group J as well).
- Scapose plants with no leafy stem; leaves all arise from a basal rosette at the sediment surface.
- Leaves with a broad blade, or compound and resembling a clover leaf in *Marsilea vestita*. - Group E
- Leaves without a broad blade.
- Leaves very long and thin; limp. - Group F
- Leaves short, often fleshy or terete; stiff. - Group G
- Leafy stems with leaves scattered throughout the water column.
- Leaves all finely dissected, (in the Characeae there are no leaves, the analogous structures are branches and branchlets which are repeatedly forked). - Group H
- Leaves entire.
- Leaves filiform, long and narrow with no obvious blade. - Group I
- Leaves more or less broad with an obvious blade or wider portion. - Group J
- Vegetative parts may be floating or emergent from the water. (in the spring immature plants may not have reached the waters surface; late season flooding may inundate plants that were previously floating on the surface).

- Floating leaves present. (submersed leaves may also be present but generally no emergent leaves are present).
- Only floating leaves are present. - Group K
- Submersed and floating leaves are present.
- Leafy stems beneath the surface of the water with distinct floating leaves on the surface as well (the floating leaves of *Potamogeton alpinus* may not be distinctly different from the submersed leaves). - Group L
- Long, thin, limp, leaves arise beneath the surface of the water and the upper ends of these leaves float on the surface of the water. - Group M
- Leaves and/or stems emergent into the air. (generally no floating leaves are present except in *Sparganium* but some floating-leaf plants with stiff petioles may resemble Group O plants if water levels drop and the floating leaves are left emergent; *Nuphar* is an example).
- Leaves and stems lax and decumbent; sprawling over the surface of the water or adjacent shore. - Group N
- Leaves and stems rigid, erect and emergent. These are generally marginal plants of shallow water and may grow completely out of the water in late summer when water levels drop. - Group O

The Species Found in Each Growth-form Group

Group	*Genera and Species*	*Genera and Species*
A	*{Eichhornia crassipes}*	*{Limnobium laevigatum}*
	{Limnobium spongia}	*{Pistia stratiotes}*
	{Salvinia}	*{Trapa natans}*
B	*Azolla caroliniana*	*Azolla filiculoides*
	Azolla mexicana	*Lemna minor*
	Riccia fluitans	*Ricciocarpus natans*
	Spirodela polyrhizas	*Wolffia borealis*
	Wolffia columbiana	*{Wolffiella floridana}*
C	*Ceratophyllum demersum*	*Lemna trisulca*
	Utricularia gibba	*Utricularia minor*
	Utricularia vulgaris	*Wolffia borealis*
	Wolffia columbiana	*{Wolffiella floridana}*
D	*Lemna trisulca*	*Utricularia gibba*
	Utricularia minor	*Utricularia vulgaris*

Contd...

Group	Genera and Species	Genera and Species
E	*Limosella aquatica*	*Marsilea vestita*
	Ranunculus flammula	*Ranunculus cymbalaria*
F	*Sagittaria*	*Vallisneria americana*
	Sparganium resemble this group	but are not scapose; they have leafy stems
G	*Isoetes*	*Lilaeopsis occidentalis*
	Lobelia dortmanna	{*Pilularia americana*}
	Ranunculus flammula	
H	{*Cabomba caroliniana*}	*Ceratophyllum*
	Chara	{*Limnophila sessiliflora*}
	Megalodonta beckii	*Myriophyllum*
	Nitella	*Ranunculus aquatilis*
	Ranunculus flabellaris	*Ranunculus sceleratus*
	Tolypella intricata	*Utricularia intermedia*
I	*Callitriche*	*Chara*
	Heteranthera dubia	*Hippuris vulgaris*
	Juncus supiniformis	*Najas flexilis*
	Nitella	*Phyllospadix* (marine)
	Potamogeton berchtoldii	*Potamogeton filiformis*
	Potamogeton foliosus	*Potamogeton friesii*
	Potamogeton obtusifolius	*Potamogeton pectinatus*
	Potamogeton pusillus	*Potamogeton robbinsii*
	Potamogeton strictifolius	*Potamogeton vaginatus*
	Potamogeton zosteriformis	*Ruppia maritima*
	Subularia aquatica	*Tolypella intricata*
	Zannichellia palustris	*Zostera* (marine)
J	*Crassula aquatica*	*Elatine rubella*
	{*Egeria densa*}	*Elodea*
	{*Hydrilla verticillata*}	{*Lagarosiphon major*}
	Ludwigia palustris	*Potamogeton crispus*
K	*Brasenia schreberi*	*Caltha natans*
	Nuphari	*Nymphaea*
	Nymphoides	*Polygonum amphibium*
	Potamogeton natans	*Ranunculus hyperboreus*
	Ranunculus lobbii	

Contd...

Group	Genera and Species	Genera and Species
L	*Callitriche*	*Potamogeton alpinus*
	Potamogeton amplifolius	*Potamogeton epihydrus*
	Potamogeton gramineus	*Potamogeton illinoensis*
	Potamogeton natans	*Potamogeton nodosus*
	Potamogeton oakesianus	*Potamogeton perfoliatus*
	Potamogeton praelongus	*Potamogeton richardsonii*
	Ranunculus aquatilis	*Ranunculus hyperboreus*
	Ranunculus lobbii	
M	*Glyceria*	*Sagittaria*
	Scirpus subterminalis	*Sparganium*
	Vallisneria americana	
N	*Callitriche*	*Heteranthera dubia*
	Juncus supiniformis	*Ludwigia palustris*
	Mimulus	*Myosotis*
	{Myriophyllum aquaticum}	*Nasturtium officinale*
	Veronica	
O	All the rooted, emergent plants belong to this group.	

Keys to the Species of the Growth-form Groups

Group A

- plants with long stems and leaves distributed along the stems. - {*Salvinia*}
- plants with all leaves in a compact rosette arising from a common point.
- petioles inflated to serve as floats.
- leaf bases cordate or abruptly narrowed to the petiole, flowers large, showy and purple, leaves glabrous and margins entire, rounded in outline, parallel veined. - {*Eichhornia crassipes*}
- leaf bases cuneate to truncate, flowers short-petiolate and inconspicuous, leaves rhombic and angular, apical margins at least toothed. - {*Trapa natans*}
- petioles not inflated.
- leaves petiolate, glabrous or nearly so, round to heart-shaped with a basal notch, spongy, green or marked with red, flat on the surface of the water. - {*Limnobium*}

- leaves have a thick spongy layer and are arched on the top surface, petiole short and leaves therefore closely clustered, leaves not marked. - {*Limnobium laevigatum*}
- leaves with a thin spongy layer, flat on the dorsal surface, petioles longer than the blades which are remote, leaves with dorsal red-brown markings. - {*Limnobium spongia*}
- leaves sessile or sub-sessile, pubescent, obovate and apically notched, leathery and grey-green, longitudinally ribbed., erect and emergent. - {*Pistia stratiotes*}

Group B

A compound microscope is required to positively identify species of *Azolla*.

- plants reduced to thalli with no stems or leaves, roots may be present.
- thalli not repeatedly and regularly forked, elongate to orbicular, flat or globose, flowering plants (duckweeds). - Lemnaceae
- one or more roots and nerves on each thallus (frond).
- each thallus bears two or more clustered roots from the base and 4 to 12 nerves. - *Spirodela polyrhiza*
- each thallus bears one root at the base and 1 to 3 nerves. - *Lemna*
- fronds oblong to lanceolate, 6 to 12 mm long and connected in small groups by stalks of the same length, matted, generally submersed, colonies. - *Lemna trisulca*
- fronds oval to round, less than 6 mm long, no stalks, solitary or in small attached groups, floating on the surface, usually 1 nerved, 2 to 4 mm long. - *Lemna minor*
- no roots or nerves on the thalli.
- thalli sub-globose to oval, up to 1 mm long. - *Wolffia*
- plants floating just below the surface, sub-globose, upper surface rounded, green, not puncticulate, 0.5 to 1.0 mm long. - *Wolffia columbiana*
- plants floating on the surface, ellipsoidal, upper surface flattish, white- or brown-punctilate, 0.5 to 1.2 mm long and about 1/2 as wide. - *Wolffia borealis*
- thalli long and narrow, sickle-shaped, several mm long but very narrow, flattened, submersed except at the base and aggregated into clusters. - {*Wolffiella floridana*}

- thalli repeatedly and regularly forked, bryophytes (liverworts).
- thalli broad, 2 to 3 times forked, tinged with purple, and fringed with tongue-shaped toothed scales. - *Ricciocarpus natans*
- thalli narrow and elongate, repeatedly dichotomously forked, not purple tinged nor fringed with scales. - *Riccia fluitans*
- stems with leaves and roots, leaves scale-like and overlapping, often reddish, ferns. - *Azolla*
- glochidia with several (usually 3 or more) septa, leaves at least 0.7 mm long, plant 1 to 3 cm in diameter, submersed leaf-lobes much larger than the upper lobes, glochidia unbranched. - *Azolla mexicana*
- glochidia with 1 (2) or no septa.
- glochidia without any septa, leaves papillose, oblong to ovate, about 1 mm long, megaspores coarsely roughened, plants 1 to 10 cm long, branches numerous and open, lower submersed leaf-lobe about as large as the upper lobe, glochidia unbranched. - *Azolla filiculoides*
- glochidia with 1 or 2 apical septa, leaves nearly smooth, sub-orbicular, about 0.5 mm long, megaspores finely roughened, plants usually less than 3 cm long, few branches and crowded, lower submersed leaf-lobe glabrous, larger and paler than the upper lobe, glochidia may be branched. - *Azolla caroliniana*

Group C

- plants with roots, stems and dissected leaves.
- plants with bladders on the alternate leaves. - *Utricularia*
- leaves divided into ultimately terete or threadlike segments.
- leaves divided into fewer than 5 final threadlike segments, leaf margins glabrous, bladders scarce on a small delicate plant usually floating at the surface or entangled in other rooted plants. - *Utricularia gibba*
- leaves 'pinnatifid', more than 20 terete final segments, hairy leaf margins, many bladders on the ordinary leaves, a robust plant to several meters long, usually lying on the sediment surface. - *Utricularia vulgaris*
- leaves divided di- or trichotomously into ultimately flattened segments.
- ordinary leaves generally with a few bladders, leaf margins glabrous, the terminal leaf segments are acuminate. - *Utricularia minor*

- ordinary leaves rarely, if ever, with bladders which are found on separate subterranean branches, leaf margins hairy, the ultimate leaf segment tips are awned. - *Utricularia intermedia*
- plants without bladders on the whorled leaves. - *Ceratophyllum*
- leaf segments sub-capillary, mostly entire, delicate and light green, in deeper water and not surfacing, achene with 3-5 lateral spines on each side, not a 'weedy' species. - *Ceratophyllum echinatum*
- leaf segments capillary to linear and flattened, serrate to coarsely toothed, plant usually coarse and robust, dark green to almost black, usually surfacing, achene without lateral spines, 2 basal spines and 1 terminal spine only, a very 'weedy' species in eutrophic waters. - *Ceratophyllum demersum*
- plants reduced to thalli with no stems or leaves, roots may be present. - Lemnaceae
- one or more roots and nerves on each thallus (frond).
- each thallus bears two or more clustered roots from the base and 4 to 12 nerves. - *Spirodela polyrhiza*
- each thallus bears one root at the base and 1 to 3 nerves. - *Lemna*
- fronds oblong to lanceolate, 6 to 12 mm long and connected in small groups by stalks of the same length, matted, generally submersed, colonies. - *Lemna trisulca*
- fronds oval to round, less than 6 mm long, no stalks, solitary or in small attached groups, floating on the surface, usually 1 nerved, 2 to 4 mm long. - *Lemna minor*
- no roots or nerves on the thalli.
- thalli sub-globose to oval, up to 1 mm long. - *Wolffia*
- plants floating just below the surface, sub-globose, upper surface rounded, green, not puncticulate, 0.5 to 1.0 mm long. - *Wolffia columbiana*
- plants floating on the surface, ellipsoidal, upper surface flattish, white- or brown-punctilate, 0.5 to 1.2 mm long and about 1/2 as wide. - *Wolffia borealis*
- thalli long and narrow, sickle-shaped, several mm long but very narrow, flattened, submersed except at the base and aggregated into clusters. - {*Wolffiella floridana*}

Group D

- plants reduced to thalli with no stems or leaves, roots may be present. - Lemnaceae
- one or more roots and nerves on each thallus (frond).
- each thallus bears two or more clustered roots from the base and 4 to 12 nerves. - *Spirodela polyrhiza*
- each thallus bears one root at the base and 1 to 3 nerves. - *Lemna*
- fronds oblong to lanceolate, 6 to 12 mm long and connected in small groups by stalks of the same length, matted, generally submersed, colonies. - *Lemna trisulca*
- fronds oval to round, less than 6 mm long, no stalks, solitary or in small attached groups, floating on the surface, usually 1 nerved, 2 to 4 mm long. - *Lemna minor*
- no roots or nerves on the thalli.
- thalli sub-globose to oval, up to 1 mm long. - *Wolffia*
- plants floating just below the surface, sub-globose, upper surface rounded, green, not puncticulate, 0.5 to 1.0 mm long. - *Wolffia columbiana*
- plants floating on the surface, ellipsoidal, upper surface flattish, white- or brown-punctilate, 0.5 to 1.2 mm long and about 1/2 as wide. - *Wolffia borealis*
- thalli long and narrow, sickle-shaped, several mm long but very narrow, flattened, submersed except at the base and aggregated into clusters. - {*Wolffiella floridana*}
- plants with roots, stems and dissected leaves bearing bladders. - *Utricularia*
- leaves divided into ultimately terete or threadlike segments.
- leaves divided into fewer than 5 final threadlike segments, leaf margins glabrous, bladders scarce on a small delicate plant usually floating at the surface or entangled in other rooted plants. - *Utricularia gibba*
- leaves 'pinnatifid', more than 20 terete final segments, hairy leaf margins, many bladders on the ordinary leaves, a robust plant to several meters long, usually lying on the sediment surface. - *Utricularia vulgaris*
- leaves divided di- or trichotomously into ultimately flattened segments.

- ordinary leaves generally with a few bladders, leaf margins glabrous, the terminal leaf segments are acuminate. - *Utricularia minor*
- ordinary leaves rarely, if ever, with bladders which are found on separate subterranean branches, leaf margins hairy, the ultimate leaf segment tips are awned. - *Utricularia intermedia*

Group E

- ferns with peltate, 4-lobed, 'clover-like' leaves coiled in the bud, long petioles. - *Marsilea vestita*
- flowering plants, leaves not 4-lobed and 'clover-like', leaves not coiled in the bud, short-petiolate or sessile.
- leaves broad, lobed or crenate, cordate, flowers yellow, fruit a head of achenes. - *Ranunculus cymbalaria*
- leaves entire and linear.
- flowers white, fruit a capsule. - *Limosella aquatica*
- flowers yellow, fruit a head of achenes. - *Ranunculus flammula*

Group F

Juvenile or sterile specimens may be difficult to distinguish.

- plants rhizomatous but not tuberous, flowers small and inconspicuous, on the end of a very long coiled pedicel which retracts below the water surface in fruit, flaccid leaves with many small longitudinal veinlets and cross-septa. - *Vallisneria americana*
- plants rhizomatous and tuberous, flowers white and showy in emergent racemes. - *Sagittaria*
- leaves basally sagittate or hastate, stamens 7 to 25, achenes densely packed on the receptacle, plants emergent in shallow water.
- monoecious, mature achenes 2.0 to 2.5 mm long with a beak less than 0.5 mm long pointing forward from the tip of the achene. - *Sagittaria cuneata*
- monoecious or dioecious, mature achenes (2.5) 3.0 to 3.5 (4.0) mm long with a beak about 1 mm long at right angles to the body of the achene. - *Sagittaria latifolia*
- leaves gradually tapered or cordate at the base; broadly ribbon-like, completely submersed in deep water or floating on the surface, ribbon-like or tapered at the base and oval, flaccid, plants generally sterile. - *Sagittaria*

In *Sagittaria* the usual non-achene key characters of bract length and shape and pedicel lengths have been found to be inconsistent and sufficiently variable even within one population as to be virtually useless as key characters. Many flowering, but non-fruiting, collections are difficult to identify.

Group G

A compound or dissecting microscope is required to positively identify species of *Isoetes*. Magnification over 10x is needed to examine the megaspore surface.

- ferns, leaves coiled in the bud, hairy sporocarps on the rhizome. - {*Pilularia americana*}
- not ferns, leaves not coiled in the bud, no hairy sporocarps.
- leaves not with 4 transversely septate, longitudinal gas chambers, flowering plants.
- leaves reduced to elongate, narrow, entire, hollow, transversely septate phyllodes, flowers in short umbels. - *Lilaeopsis occidentalis*
- leaves not as above, flowers not umbellate.
- plants solitary but usually colonial, a large cluster of white roots, white flowers in scapose racemes. - *Lobelia dortmanna*
- plants stoloniferous, no large conspicuous cluster of white roots, flowers on pedicels, solitary from near the ends of the stolons, flowers yellow, fruit a head of achenes. - *Ranunculus flammula*
- leaves sharply terete, sheathing bases with spores on the inside face of the base, each with 4 transversely septate, longitudinal gas chambers, stems short and tuberous. - *Isoetes*
- corms 3-lobed, plants primarily terrestrial, grow in grassy, ephemeral pools which are wet in winter and dry in summer. - *Isoetes nuttallii*
- corms 2-lobed, plants primarily of lake bottoms but sometimes exposed in summer due to water level fluctuations.
- megaspores spiny.
- megaspores vary greatly in size and are often aborted, spines blunt and dense. (this is probably a hybrid of *I. echinospora* and *I. maritima*). - *Isoetes truncata*
- megaspores uniform in size, spines sharp, not crowded.

- plants flaccid, megaspore spines elongate with pointed tips, uniform in size, microspores smooth. - *Isoetes echinospora*
- plants rigid, megaspore spines stubby and blunt, shorter and denser along the ridges, microspores spiny. - *Isoetes maritima*
- megaspores smooth or with ridges but not spiny.
- lowermost leaves in 2 ranks, megaspores over 0.5 mm in diameter, microspores rugose and over 36μm long. - *Isoetes occidentalis*
- lowermost leaves spirally arranged, megaspores less than 0.5 mm in diameter, microspore spiny and less than 30μm long.
- base of the leaves blackened, hyaline wing margins of the leaves extending 1 to 5 cm above the sporangium, sporangia brown spotted. - *Isoetes howellii*
- base of the leaves green, hyaline wing margins of the leaves not extending more than 1 cm above the sporangium, sporangia without colour. - *Isoetes bolanderi*

Group H

- flowering plants with roots, stems and leaves. - General Key: Aquatic Plants of British Columbia-Part 2.
- algae, no roots or leaves, only branches, produces clusters of reddish spores. - Characeae
- branchlets simple, not further branched or segmented, coronula 5-celled. - *Chara*
- stem not corticate, stipulodes in 1 tier, and alternating with the branchlets, branchlets tipped with a corona, no spine cells. - *Chara braunii*
- stems, and at least basal branchlet segment, corticate.
- spine cells in fascicles, axial cortex haplostichous, 1 corticate, stipulodes in 2 tiers, spiny looking, found in saline or brackish waters. - *Chara canescens*
- spine cells solitary or geminate, axial cortex diplostichous or triplostichous, 2 or 3 corticate.
- axial cortex diplostichous, stipulodes in 2 tiers, spine cells absent or small, rarely geminate, quite variable and widespread in hard-water lakes, smooth but marl encrusted. - *Chara vulgaris*
- axial cortex triplostichous, stipulodes usually in 2 tiers but may be in 1 tier, obscure or absent, spine cells usually absent or obscure but may be clustered, variable and widespread in

hard-water and soft-water lakes, smooth appearance. - *Chara globularis*

- branchlets compound, re-branched and/or segmented.
- branchlets simple but segmented like a string of sausages, end cells reduced and acute, heads coarse and like a bird's nest. - *Tolypella intricata*
- branchlets compound, re-branched, may be segmented or apparently so. - *Nitella*
- the end segment of the branchlets is 1-celled.
- heteroclemous, 2 kinds of branchlets in a whorl, 1-forked alternating with 1-celled, bright green, compact, branchlets swollen. - *Nitella clavata*
- homeoclemous, all branchlets in a whorl the same, generally 1-forked.
- fertile heads small and densely compact, no mucous, generally no terminal dactyls on the sterile branchlets. - *Nitella acuminata*
- fertile heads absent or loose and not compact.
- dactyl apices not long acuminate, may be acute, blunt or apiculate, (a small variety may have 2-forked branchlets occasionally). - *Nitella flexilis*
- dactyl apices are long acuminate, gametangia on branchlets, generally not marl covered. - *Nitella acuminata*
- the end segment of the branchlets is 2-or-more-celled, the end cells are very much smaller than the penultimate cell, branchlets furcate, heads terminal, branchlets homeoclemous.
- robust plants, over 15 cm high, axes over 600 microns in diameter, dactyls may be small. - Nitella furcata
- smaller plants, seldom over 15 cm tall, axes 200 to 450 microns in diameter, dactyls uniform.
- lowest branchlet node fertile, dactyls 2 to 3-celled, heads may have mucous, oospore membrane granular or felt-like. - *Nitella gracilis*
- lowest branchlet node sterile, dactyls 2-celled, heads without mucous, oospore membrane reticulate. - *Nitella tenuissima*

Group I

Mature fruits are generally needed for positive identification of *Callitriche.*

- algae, produce clusters of reddish spores, no flowers, roots or leaves. - Characeae
- branchlets simple, not further branched or segmented, coronula 5-celled. - *Chara*
- stem not corticate, stipulodes in 1 tier, and alternating with the branchlets, branchlets tipped with a corona, no spine cells. - *Chara braunii*
- stems, and at least basal branchlet segment, corticate.
- spine cells in fascicles, axial cortex haplostichous, 1-corticate, stipulodes in 2 tiers, spiny looking, found in saline or brackish waters. - *Chara canescens*
- spine cells solitary or geminate, axial cortex diplostichous or triplostichous, 2- or 3-corticate.
- axial cortex diplostichous, stipulodes in 2 tiers, spine cells absent or small, rarely geminate, quite variable and widespread in hard-water lakes, smooth but marl encrusted. - *Chara vulgaris*
- axial cortex triplostichous, stipulodes usually in 2 tiers but may be in 1 tier, obscure or absent, spine cells usually absent or obscure but may be clustered, variable and widespread in hard-water and soft-water lakes, smooth appearance. - *Chara globularis*
- branchlets compound, re-branched and/or segmented.
- branchlets simple but segmented like a string of sausages, end cells reduced and acute, heads coarse and like a bird's nest. - *Tolypella intricata*
- branchlets compound, re-branched, may be segmented or apparently so. - *Nitella*
- the end segment of the branchlets is 1-celled.
- heteroclemous, 2 kinds of branchlets in a whorl, 1-forked alternating with 1-celled, bright green, compact, branchlets swollen. - *Nitella clavata*
- homeoclemous, all branchlets in a whorl the same, generally 1-forked.
- fertile heads small and densely compact, no mucous, generally no terminal dactyls on the sterile branchlets. - *Nitella acuminata*
- fertile heads absent or loose and not compact.
- dactyl apices not long acuminate, may be acute, blunt or apiculate, (a small variety may have 2-forked branchlets occasionally). - *Nitella flexilis*

- dactyl apices are long acuminate, gametangia on branchlets, generally not marl covered. - *Nitella acuminata*
- the end segment of the branchlets is 2-or-more-celled, the end cells are very much smaller than the penultimate cell, branchlets furcate, heads terminal, branchlets homeoclemous.
- robust plants, over 15 cm high, axes over 600 microns in diameter, dactyls may be small. - *Nitella furcata*
- smaller plants, seldom over 15 cm tall, axes 200 to 450 microns in diameter, dactyls uniform.
- lowest branchlet node fertile, dactyls 2- to 3-celled, heads may have mucous, oospore membrane granular or felt-like. - *Nitella gracilis*
- lowest branchlet node sterile, dactyls 2-celled, heads without mucous, oospore membrane reticulate. - *Nitella tenuissima*
- flowering plants with roots, stems and leaves.
- marine plants. - Zosteraceae
- leaf sheaths deciduous, sometimes leaving a few scaly parts behind, leaf blades thin and translucent, rhizome with elongate internodes (1 to 3 cm long or more), with 2 thin roots at each internode, monoecious, spadix border projections inconspicuous, if present, usually established on sand or mud. - *Zostera*
- leaves 3-veined, 1 to 1.5 mm wide, sheaths split to the base, rare and probably introduced from Asia, known from Boundary Bay and Tsawwassen. - *Zostera japonica*
- leaves 5-veined or more, 1.5 to 12 mm wide, sheaths on sterile shoots are closed at the base, widespread native species. - *Zostera marina*
- basal portions of the leaf sheaths decay with age to bundles of fine, woolly fibres, leaf blades leathery, rhizome has short, thick internodes with 2 or more thick roots at each internode, dioecious, spadix bordered by conspicuous flap-like projections, usually established only on hard or rocky substrates. - *Phyllospadix*
- fertile stems branched, 4 to 12 dm long, spathes usually paired at the nodes. - *Phyllospadix torreyi*
- fertile stems unbranched, 0.5 to 4 dm long, spathes usually solitary at the nodes.
- leaves with 3 (rarely 5) veins, margins entire. - *Phyllospadix scouleri*

- leaves with 5 or 7 veins, margins toothed towards the apex. - *Phyllospadix serrulatus*
- fresh water or brackish water plants.
- inflorescence an umbel on a long coiled pedicel, brackish water. - *Ruppia maritima*
- inflorescence not umbellate, pedicel not coiled in fruit.
- leaves in a basal cluster, flowers racemose. - *Subularia aquatica*
- leaves cauline.
- leaves whorled or opposite.
- leaves in whorls of 6 or more, submersed leaves to 5 cm long or more, flowers axillary, stem apex often emergent. - *Hippuris vulgaris*
- leaves opposite, submersed leaves 2 (2.5) cm long, flowers axillary, stem apices may form a rosette floating on the surface. - *Callitriche*
- fruit encircled by a conspicuous wing-like margin, leaf bases joined by winged ridges. - *Callitriche stagnalis*
- fruit not winged or with only a narrow wing at the tip, leaf bases various.
- leaves all linear, 1-nerved, light green, leaf bases not joined by a wing floral bracts absent, common species. - *Callitriche hermaphroditica*
- leaves various, upper often ovate and 3-nerved, bases joined by a wing-like ridge, floral bracts present.
- carpel face markings in regular vertical lines, fruit slightly wing-margined at the top and longer than broad. - *Callitriche verna*
- carpel face markings scattered, fruit not winged and as long as broad.
- fruits widest above the middle (obovate) leaves bidentate, midvein barely thickened at the end, emergent leaves may be over 5 mm wide, stems long. - *Callitriche heterophylla*
- fruits round or oblong, midvein thickened at the tip and protruding, leaves never over 5 mm wide, plants short and slender. - *Callitriche anceps*
- leaves alternate.
- plants stoloniferous and proliferous, forming a dense, often reddish, tangled mat on the surface, leaves auriculate, perianth brown. - *Juncus supiniformis*

- plants not stoloniferous, proliferous or mat-forming, leaves not auriculate.
- fruits a cluster of banana-shaped achenes in the leaf axils, female flowers in a cup-shaped sheath. - *Zannichellia palustris*
- fruits not a cluster of banana-shaped achenes, flowers not in cup-shaped sheaths.
- plants sprawling and rooted at the lower nodes at least, flower pedicels arising from the stipulate leaf axils, solitary yellow flowers from spathes, fruit a capsule. - *Heteranthera dubia*
- plants erect, not rooting at the nodes, flowers small, numerous and inconspicuous in terminal, pedunculate spikes, no spathes, fruit an achene. - *Potamogeton*

Group J

- leaves alternate.
- leaves sessile and clasping, crispate, finely serrulate, leaves oblong 3 to 8 cm long and 3 to 10 mm wide, flowers in a terminal emergent inflorescence. - *Potamogeton crispus*
- leaves not clasping or crispate, inflorescence axillary.
- leaves strongly recurved, serrate or toothed, sessile, linear to linear-lanceolate. - {*Lagarosiphon major*}
- leaves straight, entire, petiolate, ovate-elliptic. - *Ludwigia palustris*
- leaves opposite or whorled.
- leaves opposite, entire, often fleshy, flowers solitary in the leaf axils.
- flowers 4-merous, 4 erect purplish follicles with 6 to 12 seeds. - *Crassula aquatica*
- flowers 2- or 3-merous, capsules with extensively pitted, curved seeds. - *Elatine rubella*
- leaves whorled, may be toothed, membranous. - *Hydrocharitaceae*
- leaves very long and narrow in basal rosettes, pedicel arises from the rosette and reaches the surface, spirally coiled. - *Vallisneria americanus*
- leaves short, up to 4 cm long, arising from the elongated stem.
- leaf arrangement irregular, margins serrate or toothed, tip with 2 enlarged spines - {*Lagarosiphon major*}

- leaves arranged in regular whorls.
- leaf margins serrate or toothed.
- leaf tip with 2 enlarged spines, no turions, 3 or more strongly recurved leaves up to 3 cm long per whorl. - {*Lagarosiphon major*}
- no enlarged spines on the leaf tip, turions present on the rhizome or on the stem, 3 to 8 (12) nearly straight leaves up to 2.5 cm long per whorl. - {*Hydrilla verticillata*}
- leaf margins entire.
- basal leaves in whorls of 3 but upper ones up to 6 per whorl and to 4 cm long, 2 to 3 flowers in the staminate spathes, petals to 10 mm. - {*Egeria densa*}
- basal leaves in pairs, the upper ones usually in whorls of 3 and up to 2.5 cm long, flowers solitary in the staminate spathes, petals up to 5 mm long. - *Elodea*
- leaves in whorls of 3.
- leaves rarely over 1.5 cm long.
- leaves 2 (1 to 4) mm wide, tapered abruptly to a blunt point, staminate flowers stalked and persistent. - *Elodea canadensis*
- leaves 1.5 (0.3 to 1.5) mm wide, tapered to a slender point, staminate flowers sessile and deciduous at anthesis. - *Elodea nuttallii*
- leaves (1.7) 2.0 to 2.6 cm long. - {*Elodea longivaginata*}
- leaves in the upper and middle part of the stem in pairs, lower ones irregular or alternate, (1.7) 2.0 to 2.6 cm long. - {*Elodea longivaginata*}

Group K

- leaves peltate, submersed portions of the plants mucilaginous. - *Brasenia schreberi*
- leaves not peltate, petiole attached at the leaf margin, plants not mucilaginous.
- leaves entire, over 1 cm long, flowers not yellow.
- leaves orbicular, bases hastate or cordate.
- fruit is a berry, 4 to 6 green sepals, many white, yellow or red petals and many stamens, rhizome massive. - Nymphaeaceae
- fruits a cluster of follicles, 5 or 6 small white sepals, no petals and up to 20 stamens, slender creeping stolons. - *Caltha natans*

- leaves not orbicular, longer than broad, bases cordate, reniform or tapered.
- leaves cordate or reniform at the base, ovate to elliptic.
- leaves acute at the apex and cordate at the base, usually reddish, petioles reflexed at the leaf junction, leaf margin entire. - *Potamogeton natans*
- leaves rounded at the apex and reniform to cordate at the base, greenish, leaf margin toothed. - *Caltha natans*
- leaves obtuse to basally tapered, lanceolate to narrowly-elliptic, apex acute, green, flowers rose or pink in compact panicles. - *Polygonum amphibium*
- leaves lobed, up to about 1 cm long, flowers yellow, fruit a head of achenes. - *Ranunculus hyperboreus*

Group L

Mature fruits are generally needed for positive identification of *Callitriche.*

- leaves opposite, entire, 1 or 3 nerved, up to 2 (2.5) cm long, no stipules, flowers sessile in the axils and submersed. - *Callitriche*
- fruit encircled by a conspicuous wing-like margin, leaf bases joined by winged ridges. - *Callitriche stagnalis*
- fruit not winged or with only a narrow wing at the tip, leaf bases various.
- leaves all linear, 1-nerved, light green, leaf bases not joined by a wing, floral bracts absent, common species. - *Callitriche hermaphroditica*
- leaves various, upper often ovate and 3-nerved, bases joined by a wing-like ridge, floral bracts present.
- carpel face markings in regular vertical lines, fruit slightly wing-margined at the top and longer than broad. - *Callitriche verna*
- carpel face markings scattered, fruit not winged and as long as broad.
- fruits widest above the middle (obovate) leaves bidentate, midvein barely thickened at the end, emergent leaves may be over 5 mm wide, stems long. - *Callitriche heterophylla*
- fruits round or oblong, midvein thickened at the tip and protruding, leaves never over 5 mm wide, plants short and slender. - *Callitriche anceps*

- leaves alternate, entire or lobed, usually more than 3 veins, stipulate or with an inflated petiole base, flowers pedicellate, emergent.
- floating leaves entire, flowers inconspicuous in terminal, emergent inflorescences, leaves generally well over 2 cm long, stipules present. - *Potamogeton*
- floating leaves lobed, flowers solitary, axillary, yellow or white, leaves from 1 to 2 cm long, base of petiole inflated and sheathing. - *Ranunculus*
- flowers white, submersed leaves dissected into ultimately filiform segments.
- two to 7 glabrous, beakless achenes, plants and receptacle glabrous, pedicels in the axils of ternately lobed floating leaves, leaves 2 to 3 times divided into 8 to 12 segments. - *Ranunculus lobbii*
- 10 to 80 glabrous or hirsute achenes, plants and receptacles hirsute.
- 30 to 80 achenes, flowers may have yellow bases, leaf blades sessile on the stipular base. - *Ranunculus circinatus*
- ten to 25 achenes, flowers all white, leaf blades petiolate. - *Ranunculus aquatilis*
- flowers yellow, submersed leaves, if present, simple to ternately divided.
- leaves simple and entire to serrate, may be shallowly lobed, achenes glabrous.
- achenes 50 to 200 in a columnar head, longitudinally ribbed. - *Ranunculus cymbalaria*
- achenes 5 to 60 in a globular head, not longitudinally ribbed, stems decumbent to prostrate, rooting at the nodes, nectary scales usually broader than long. - *Ranunculus flammula*
- leaves all, or in part, lobed, parted or ternately dissected to filiform segments, achenes pubescent.
- annuals, erect, not nodally rooting, achene beaks inconspicuous. - *Ranunculus sceleratus*
- perennials, floating or reclining, nodally rooting, achene beaks conspicuous.
- leaves deeply lobed or parted, achenes not corky margined.

- leaves deeply 3 parted with narrow lobes distally acute, achene beaks 1/4 the length of the achenes, receptacles hirsute, 5 sepals and petals. - *Ranunculus gmelinii*
- leaves 3-lobed, distally rounded, achene beaks 1/10 the length of the achenes, receptacles glabrous, 3 sepals and petals. - Ranunculus hyperboreus
- leaves, at least the submersed ones, 3 to 5 times ternately dissected into filiform segments less than 2 mm wide, achenes corky margined. - *Ranunculus flabellaris*

Group M

- leaves filiform or terete, usually under 3 mm wide, a solitary sub-terminal spikelet in the axil of a prominent bract. - *Scirpus subterminalis*
- leaves usually broader than 3 mm, not filiform or terete.
- swollen nodes, closed sheathing leaf bases, prominent ligules, hollow culms, articulated panicles, leaves often wavy-margined on the surface of the water, grasses. - *Glyceria*
- nodes not swollen, no closed sheathing leaf base or ligules, stems not hollow, inflorescence not an articulated panicle.
- leaves thin and flat, 3 to 10 mm wide, with numerous longitudinal veinlets and cross septa, pistillate flowers solitary on the end of a long pedicel which coils and retracts beneath the surface in fruit. - *Vallisneria americana*
- flowers in emergent inflorescences.
- individual flowers numerous and inconspicuous in heads, fruit of achenes. - *Sparganium*
- individual flowers few with showy petals in whorled bracteate racemes. - *Sagittaria*
- leaves basally sagittate or hastate, stamens 7 to 25, achenes densely packed on the receptacle, plants emergent in shallow water.
- monoecious, mature achenes 2.0 to 2.5 mm long with a beak less than 0.5 mm long pointing forward from the tip of the achene. - *Sagittaria cuneata*
- monoecious or dioecious, mature achenes (2.5) 3.0 to 3.5 (4.0) mm long with a beak about 1 mm long at right angles to the body of the achene. - *Sagittaria latifolia*
- leaves gradually tapered or cordate at the base; broadly ribbon-like, submersed in deep water or floating on the surface, ribbon-

like or tapered at the base and oval, flaccid, plants generally sterile. - *Sagittaria*

Group N

Mature fruits are generally needed for positive identification of *Callitriche.*

- leaves finely dissected, pale yellowish-green, flowers small, in axillary whorls. - {*Myriophyllum aquaticum*}
- leaves not dissected.
- leaves, at least the upper ones, compound, apical lobe much larger than the laterals, flowers white in racemes, fruit an elongate silique. - *Nasturtium officinale*
- leaves simple, fruits not siliques.
- leaves opposite.
- leaves entire, flowers sessile in the axils.
- leaves elliptic to ovate, 2 to 6 cm long, petiolate, flowers 4 merous and sepaloid, fruit a capsule. - *Ludwigia palustris*
- leaves linear or spathulate to ovate, up to 2 (2.5) cm long, flowers naked, 1 stamen, 2 styles, 2-lobed achene-like fruit, often winged. - *Callitriche*
- fruit encircled by a conspicuous wing-like margin, leaf bases joined by winged ridges. - *Callitriche stagnalis*
- fruit not winged or with only a narrow wing at the tip, leaf bases various.
- leaves all linear, 1-nerved, light green, leaf bases not joined by a wing, floral bracts absent, common species. - *Callitriche hermaphroditica*
- leaves various, upper often ovate and 3-nerved, bases joined by a wing-like ridge, floral bracts present.
- carpel face markings in regular vertical lines, fruit slightly wing-margined at the top and longer than broad. - *Callitriche verna*
- carpel face markings scattered, fruit not winged and as long as broad.
- fruits widest above the middle (obovate) leaves bidentate, midvein barely thickened at the end, emergent leaves may be over 5 mm wide, stems long. - *Callitriche heterophylla*

- fruits round or oblong, midvein thickened at the tip and protruding, leaves never over 5 mm wide, plants short and slender. - *Callitriche anceps*
- leaves dentate.
- leaves broadly ovate to obovate, coarsely toothed, flowers solitary on long pedicels in the leaf axils, flowers pink-purple or yellow, large, showy and zygomorphic. - *Mimulus*
- corollas pink-purple, stems over 3 dm tall, leaves 3 to 10 cm long. - *Mimulus lewisii*
- corollas yellow with reddish or purplish markings, perennials with the upper calyx teeth larger than the others.
- rhizomatous and often stoloniferous, flowers usually 1 to 5, corollas 2 to 4 cm long. - *Mimulus tilingii*
- stoloniferous, rarely rhizomatous, usually 5 or more flowers, corollas usually less than 2 cm long. - *Mimulus guttatus*
- leaves linear-lanceolate to ovate, finely toothed, many pedicellate flowers in axillary racemes, flowers blue, small and zygomorphic. - *Veronica*
- leaves all short-petiolate. - *Veronica beccabunga*
- leaves, at least those on the middle and upper portions of the flowering shoots, sessile.
- capsule flattened, notched, wider than high, 5 to 9 seeds per locule, 1 to 2 mm long, leaves 4 to 20 times as long as wide. - *Veronica scutellata*
- capsule turgid, scarcely notched, barely wider than high, many seeds less than 0.5 mm long, leaves 1.5 to 5 times as long as wide.
- leaves 1.5 to 3 times as long as wide, fruiting pedicels ascending, capsules higher than wide if not equal, flowers blue or violet. - *Veronica anagallis-aquatica*
- leaves 2.5 to 5 times as long as wide, fruiting pedicels spreading, capsules wider than high, flowers white to pink or pale blue. - *Veronica catenata*
- leaves alternate.
- leaves filiform, terete, auriculate, plants proliferous and often reddish, 3 or 6 stamens, flowers brown, the perianth is undifferentiated. - *Juncus supiniformis*

- leaves linear to lanceolate, not filiform, terete or auriculate, plants not proliferous or reddish, stamens 3 or 5, perianth differentiated and showy.
- linear stipulate leaves, flowers yellow, stamens 3, solitary, long-pedicellate flowers from a spathe. - *Heteranthera dubia*
- corolla limb 5 to 10 mm wide, style as long as or longer than the nutlets, stoloniferous, mostly erect and not creeping or recumbent. - *Myosotis scorpioides*

Group O

These plants are keyed out in Parts 4, 5, and 6 of the General Key.

7

Keys to Aquatic Genera and Species within Families

Key to the Alismataceae of British Columbiae

In *Sagittaria* the usual non-achene key characters of bract length and shape and pedicel lengths have been found to be inconsistent and sufficiently variable even within one population as to be virtually useless as key characters. Many flowering, but non-fruiting, collections are difficult to identify.

- leaves basally sagittate or hastate, stamens 7-25, achenes densely packed on the receptacle. - *Sagittaria*
- monoecious, mature achenes 2.0-2.5 mm long with a beak less than 0.5 mm long pointing forward from the tip of the achene. - *Sagittaria cuneata*
- monoecious or dioecious, mature achenes (2.5) 3.0-3.5 (4.0) mm long with a beak about 1 mm long at right angles to the body of the achene. - *Sagittaria latifolia*
- leaves gradually tapered or cordate at the base; broadly ribbon-like if submersed.
- leaves completely submersed in deep water or floating on the surface, ribbon-like or tapered at the base and oval, flaccid, plants generally sterile. - *Sagittaria*
- leaves emergent and erect, ovate to lanceolate, stamens 6 (9), achenes in a single whorl on the receptacle. - *Alisma*
- leaf blades lanceolate to ovate, scapes longer than the leaves (petiole plus blade), achenes centrally grooved at the top. - *Alisma plantago-aquatica*

- leaf blades narrowly lanceolate to linear, scapes shorter than the leaves (petiole plus blade), achenes 2-grooved at the tip. - *Alisma gramineum*

Key to the Aquatic Apiaceae of British Columbia

- leaves with compound blades, plants erect and rooting at lower nodes only, umbels compound, base of the stem without transverse septa, roots not tuberous-thickened, ribs of the fruit prominent, corky-thickened, leaves with defined leaflets not dissected into narrow segments, primary lateral veins of the leaflets not directed in any particular direction, plants erect, calyx teeth tiny or absent. - *Sium sauve*
- leaves with simple blades or bladeless, plants creeping and rooting at all the nodes, umbels simple.
- leaves bladeless, reduced to linear or spathulate, septate stalks, leaves and umbels clustered at the nodes. - *Lilaeopsis occidentalis*
- leaves with blades, solitary on long petioles from the nodes, umbels solitary on long peduncles from the same nodes. - *Hydrocotyle*
- leaves sub-orbicular to reniform, crenate or shallowly lobed, centrally peltate. - *Hydrocotyle verticillata*
- leaves 5 to 6 lobed nearly or quite to the middle, not peltate. - *Hydrocotyle ranunculoides*

The lower-most leaves of many marginal Umbellifers may be under water, especially in the spring and may then be finely dissected with filiform segments. *Sium sauve* is especially prone to being found completely under water in the spring with all leaves filiform-dissected. Such plants will not key out properly in the general key. *Cicuta* is poisonous, use care when cutting open tubers to check for transverse septa. Wash your hands and your knife afterwards.

Key to the Aquatic Araceae of British Columbia

- plants free-floating, leaves thick and leathery, hairy, sub-sessile and cuneate, in a rosette, tropical aquarium and garden pool plants, not yet known to overwinter outdoors in BC. - {*Pistia stratiotes*}
- plants bottom-rooted, leaves glabrous, petiolate or linear and reed-like, native, leaves bifacial with broad blades and petioles, spadix basal and stalked, distinct showy spathes, leaves ovate

and up to 30 cm long, bases cordate, spathe white and up to 6 cm long, spadix short, clavate with large red berries. - *Calla palustris*

Key to the Cabombaceae of British Columbia

- leaves peltate and entire, floating on the surface, plant mucilaginous, flowers red, native. - *Brasenia schreberi*
- leaves dissected and submersed, plants not mucilaginous, flowers with white sepals and yellow petals, aquarium plant, not yet found out of cultivation. - {*Cabomba caroliniana*}

Key to the Characeae of British Columbia

- branchlets simple, not further branched or segmented, coronula 5-celled. - *Chara*
- stem not corticate, stipulodes in 1 tier, and alternating with the branchlets, branchlets tipped with a corona, no spine cells. - *Chara braunii*
- stems, and at least basal branchlet segment, corticate.
- spine cells in fascicles, axial cortex haplostichous, 1-corticate, stipulodes in 2 tiers, spiny looking, found in saline or brackish waters. - *Chara canescens*
- spine cells solitary or geminate, axial cortex diplostichous or triplostichous, 2 or 3 corticate.
- axial cortex diplostichous, stipulodes in 2 tiers, spine cells absent or small, rarely geminate, quite variable and widespread in hard-water lakes, smooth but marl encrusted. - *Chara vulgaris*
- axial cortex triplostichous, stipulodes usually in 2 tiers but may be in 1 tier, obscure or absent, spine cells usually absent or obscure but may be clustered, variable and widespread in hard-water and soft-water lakes, smooth appearance. - *Chara globularis*
- branchlets simple but segmented like a string of sausages, end cells reduced and acute, heads coarse and like a bird's nest. - *Tolypella intricata*
- branchlets compound, re-branched, may be segmented or apparently so. - *Nitella*
- the end segment of the branchlets is 1-celled.
- heteroclemous, 2 kinds of branchlets in a whorl, 1-forked alternating with 1-celled, bright green, compact, branchlets swollen. - *Nitella clavata*

- homeoclemous, all branchlets in a whorl the same, generally 1-forked.
- fertile heads small and densely compact, no mucous, generally no terminal dactyls on the sterile branchlets. - *Nitella acuminata*
- fertile heads absent or loose and not compact.
- dactyl apices not long acuminate, may be acute, blunt or apiculate, (a small variety may have 2-forked branchlets occasionally). - *Nitella flexilis*
- dactyl apices are long acuminate, gametangia on branchlets, generally not marl covered. - *Nitella acuminata*
- the end segment of the branchlets is 2-or-more-celled, the end cells are very much smaller than the penultimate cell, branchlets furcate, heads terminal, branchlets homeoclemous.
- robust plants, over 15 cm high, axes over 600 microns in diameter, dactyls may be small. - *Nitella furcata*
- smaller plants, seldom over 15 cm tall, axes 200 to 450 microns in diameter, dactyls uniform.
- lowest branchlet node fertile, dactyls - to 3-celled, heads may have mucous, oospore membrane granular or felt-like. - *Nitella gracilis*
- lowest branchlet node sterile, dactyls 2-celled, heads without mucous, oospore membrane reticulate. - *Nitella tenuissima*

Key to the Aquatic and Wetland Cyperaceae of British Columbia

- achenes enclosed in a perigynium, flowers unisexual.
- perigynium open, margins free (not covered in this key). - *Kobresia*
- perigynium closed, a dorsal suture may be evident (not covered in this key). - *Carex*
- achenes not enclosed in a perigynium, flowers perfect.
- scales distichous, in 2 vertical ranks, perianth of 6-9 bristles, stem leafy, upper leaves with axillary inflorescences. - *Dulichium arundinaceum*
- scales of the spikelet spirally arranged.
- styles basally thickened and persistent on the achenes as a tubercle.
- several to many spikelets, 1 or 2 fertile flowers or achenes per spikelet, 10 to 12 perianth bristles (not covered in this key). - *Rhynchospora*

- spikelet solitary, several to many fertile flowers or achenes per spikelet, up to 6 perianth bristles (not covered in this key). - *Eleocharis*
- styles not thickened, achenes may be apiculate.
- achenes subtended by more than 10 conspicuous and elongate white bristles, stigmas trifid and achenes trigonal (not covered in this key). - *Eriophorum*
- achenes subtended by 8 or fewer inconspicuous perianth bristles, stigmas bifid or trifid, achenes lenticular or trigonal.
- Spikelets with slightly modified lower scales, not leafy.
- Lower scales of the spikelets short, not more than half as long as the spikelet (not covered in this key). - *Eleocharis*
- Lower scales of the spikelets long, nearly as long or longer than the spikelets (not covered in this key. - *Trichophorum*
- Spikelets with subtending leafy bracts. (bristles present and/or achenes apiculate, perennials, involucral bract solitary, not leafy or spreading, apparently an extension of the culm.) - *Scirpus*
- spikelets numerous in branching inflorescences, often sessile in clusters at the ends of branches, species complex. - *Scirpus lacustris*
- spikelets solitary or few, sessile in a sessile cluster, spikelets solitary or rarely 2, stems flaccid and the apices floating on the surface of the water, leaves and stems filiform. - *Scirpus subterminalis*

Key to the Hydrocharitaceae of British Columbia

- leaves very long and narrow in basal rosettes, pedicel arises from the rosette and reaches the surface, spirally coiled - *Vallisneria americanus*
- leaves short, up to 4 cm long, arising from the elongated stem.
- leaf arrangement irregular, margins serrate or toothed, tip with 2 enlarged spines - {*Lagarosiphon major*}
- leaves arranged in regular whorls.
- leaf margins serrate or toothed.
- leaf tip with 2 enlarged spines, no turions, 3 or more strongly recurved leaves up to 3 cm long per whorl. - {*Lagarosiphon major*}

- no enlarged spines on the leaf tip, turions present on the rhizome or on the stem, 3 to 8 (12) nearly straight leaves up to 2.5 cm long per whorl. - {*Hydrilla verticillata*}
- leaf margins entire.
- basal leaves in whorls of 3 but upper ones up to 6 per whorl and up to 4 cm long, 2 to 3 flowers in the staminate spathes, petals to 10 mm.{*Egeria densa*}
- basal leaves in pairs, the upper ones usually in whorls of 3 and up to 2.5 cm long, flowers solitary in the staminate spathes, petals up to 5 mm long. - *Elodea*
- leaves in whorls of 3.
- leaves rarely over 1.5 cm long.
- leaves 2 (1 to 4) mm wide, tapered abruptly to a blunt point, staminate flowers stalked and persistent. - *Elodea canadensis*
- leaves 1.5 (0.3 to 1.5) mm wide, tapered to a slender point, staminate flowers sessile and deciduous at anthesis. - *Elodea nuttallii*
- leaves (1.7) 2.0 to 2.6 cm long. - {*Elodea longivaginata*}
- leaves in the upper and middle part of the stem in pairs, lower ones irregular or alternate, (1.7) 2.0 to 2.6 cm long. - {*Elodea longivaginata*}

Key to the Lemnaceae of British Columbia

- one or more roots and nerves on each thallus (frond).
- each thallus bears two or more clustered roots from the base and 4 to 12 nerves. - *Spirodela polyrhiza*
- each thallus bears one root at the base and 1 to 3 nerves. - *Lemna*
- fronds oblong to lanceolate, 6 to 12 mm long and connected in small groups by stalks of the same length, matted, generally submersed, colonies. - *Lemna trisulca*
- fronds oval to round, less than 6 mm long, no stalks, solitary or in small attached groups, floating on the surface, usually 1 nerved, 2 to 4 mm long. - *Lemna minor*
- no roots or nerves on the thalli.
- thalli sub-globose to oval, up to 1 mm long. - *Wolffia*
- plants floating just below the surface, sub-globose, upper surface rounded, green, not puncticulate, 0.5 to 1.0 mm long. - *Wolffia columbiana*

- plants floating on the surface, ellipsoidal, upper surface flattish, white- or brown-punctilate, 0.5 to 1.2 mm long and about 1/ 2 as wide. - *Wolffia borealis*
- thalli long and narrow, sickle-shaped, several mm long but very narrow, flattened, submersed except at the base and aggregated into clusters. - {*Wolffiella floridana*}

Key to the Menyanthaceae of British Columbia

- leaves trifoliate and usually entire, corolla with a mass of fimbriate scales, spicate. - *Menyanthes trifoliata*
- leaves simple, hastate or cordate, the white or yellow flowers are either solitary or clustered and appearing to arise from the petiole or base of the leaf, leaves small to about 15 cm long. - *Nymphoides*
- flowers yellow, solitary and axillary. - {*Nymphoides peltata*}
- flowers white, clustered at the base of the leaf.
- leaf blades ovate and mostly under 5 cm long, not dark-punctate beneath, seeds smooth. {*Nymphoides cordatum*}
- leaf blades orbicular and 8 to 15 cm in diameter, dark-punctate or pitted beneath, seeds glandular, warty. - {*Nymphoides aquatica*}

Key to the Nymphaeaceae of British Columbia

- large discoid stigma, superior ovary, about 6 large yellowish conspicuous sepals, flowers yellow, leaves flat, dark green, entire margined and leathery, native, large robust plants. - *Nuphar*
- stamens reddish, petiole round in cross section, primarily western and coastal BC. - *Nuphar polysepalum*
- stamens yellow, petiole flattened in cross section, primarily eastern BC. - *Nuphar variegatum*
- several to many spreading stigmas, partially inferior ovary, 4 greenish inconspicuous sepals, flowers red, pink, white and yellow, leaves often suffused with red especially below and often wavy margined but not leathery, many horticultural varieties are naturalized and cause confusion. - *Nymphaea*
- petals yellow or red, pink or reddish.
- about 25 yellow petals, flowers 6 to 10 cm in diameter. - *Nymphaea mexicana*

- petals red, pink, reddish (horticultural varieties). - *Nymphaea*
- petals white.
- flowers 7 to 12 cm in diameter, over 20 petals, over 8 styles.
- white petals gradually transitional to the staminodia and stamens, spirally arranged, stigmatic disc with 8 to 24 radial stigmas, not fragrant. - *Nymphaea alba*
- 20 to 30 thick white elliptic petals followed by 70 to 100 yellow stamens and staminodia, about 20 styles, flowers very fragrant, open in the early morning and close later in the day. - *Nymphaea odorata*
- carpellary appendages 3 mm long or more, flaccid and purplish. - *Nymphaea tetragona*
- carpellary appendages 1.5 mm long or less, stiff and greenish. - *Nymphaea leibergii*

Key to the Pontederiaceae of British Columbia

- plants erect and emergent, fruit a 1-seeded nut enclosed by the base of the perianth tube, locally introduced and usually in cultivation, 6 stamens, inflorescence geniculate at anthesis, flowers blue. - {*Pontederia cordata*}
- plants submersed, sprawling or floating on the surface, fruit a 3-locular capsule.
- floating on the surface, petiole inflated, conspicuous purple flowers, 6 stamens, introduced in garden pools but not surviving outdoors over winter. - {*Eichhornia crassipes*}
- submersed or sprawling at the surface, petiole not inflated, inconspicuous yellow flowers, 3 stamens, native. - *Heteranthera dubia*

Key to the Aquatic Ranunculaceae of British Columbia

- leaves dissected, compound or lobed, petals white or yellow, sepals green and often deciduous, fruit a head of achenes. - *Ranunculus*
- leaves simple, petals absent, sepals showy and white or yellow, fruit a cluster of elongate follicles, leaves finely crenate margined. - *Caltha*
- stems stout, erect or sprawling, leaves erect, over 5 cm wide, flowers yellow. - *Caltha palustris*
- stems slender, creeping or floating, leaves floating, under 5 cm wide, flowers white. - *Caltha natans*

Key to the Aquatic Scrophulariaceae of British Columbia

- petal tube very short and the lobes apparently free, flat and spreading, lower lobe smaller than the others, 2 stamens which spread laterally. - *Veronica*
- leaves all short-petiolate. - *Veronica beccabunga*
- leaves on the middle and upper portions of the flowering shoots sessile.
- capsule flattened, notched, wider than high, 5 to 9 seeds per locule, 1 to 2 mm long, leaves 4 to 20 times as long as wide. - *Veronica scutellata*
- capsule turgid, scarcely notched, barely wider than high, many seeds less than 0.5 mm long, leaves 1.5 to 5 times as long as wide.
- leaves 1.5 to 3 times as long as wide, fruiting pedicels ascending, capsules higher than wide if not equal, flowers blue or violet. - *Veronica anagallis-aquatica*
- leaves 2.5 to 5 times as long as wide, fruiting pedicels spreading, capsules wider than high, flowers white to pink or pale blue. - *Veronica catenata*
- petal tube long or short but lobes not apparently free and all the same size, two, 4 or 5 erect stamens.
- stamen connective developed into a wide flap partially surrounding the anthers. - *Gratiola*
- pedicels with a pair of sepaloid bracteoles and 5 sepals at the summit. - *Gratiola neglecta*
- pedicels without bracteoles, only 5 sepals at the summit. - *Gratiola ebracteata*
- stamen connective not developed into a flap, sepal tube not 5-angled, no hairy patches on the abaxial petal lip.
- petal tube not 2-lipped, lobes equal or nearly so, leaves not finely dissected., with a petiole and a blade. - *Limosella aquatica*
- petal tube 2-lipped, lobes unequal, leaves finely dissected. - {*Limnophila sessiliflora*}

Key to the Umbelliferae of British Columbia

Key to the Zosteraceae of British Columbia

These are all marine plants.

- leaf sheaths deciduous, sometimes leaving a few scaly parts behind, leaf blades thin and translucent, rhizome with elongate internodes (1 to 3 cm long or more), with 2 thin roots at each internode, monoecious, spadix border projections inconspicuous, if present, usually established on sand or mud. - *Zostera*
- leaves 3-veined, 1 to 1.5 mm wide, sheaths split to the base, rare and probably introduced from Asia, known from Boundary Bay and Tsawwassen. - *Zostera japonica*
- leaves 5-veined or more, 1.5 to 12 mm wide, sheaths on sterile shoots are closed at the base, widespread native species. - *Zostera marina*
- basal portions of the leaf sheaths decay with age to bundles of fine, woolly fibres, leaf blades leathery, rhizome has short, thick internodes with 2 or more thick roots at each internode, dioecious, spadix bordered by conspicuous flap-like projections, usually established only on hard or rocky substrates. - *Phyllospadix*
- fertile stems branched, 4 to 12 dm long, spathes usually paired at the nodes. - *Phyllospadix torreyi*
- fertile stems unbranched, 0.5 to 4 dm long, spathes usually solitary at the nodes.
- leaves with 3, rarely 5, veins, margins entire. - *Phyllospadix scouleri*
- leaves with 5 or 7 veins, margins toothed towards the apex. - *Phyllospadix serrulatus*

Keys to the Aquatic Species within Genera

Key to the Alisma *of British Columbia*

- leaf blades lanceolate to ovate, scapes longer than the leaves (petiole plus blade), achenes centrally grooved at the top. - *Alisma plantago-aquatica*
- leaf blades narrowly lanceolate to linear, scapes shorter than the leaves (petiole plus blade), achenes 2-grooved at the tip. - Alisma gramineum

Key to the Azolla *of British Columbia*

A compound microscope is required to positively identify species in *Azolla*.

- glochidia with several (usually 3 or more) septa, leaves at least 0.7 mm long, plant 1 to 3 cm in diameter, submersed leaf-lobes much larger than the upper lobes, glochidia unbranched. - *Azolla mexicana*

- glochidia without any septa, leaves papillose, oblong to ovate, about 1 mm long, megaspores coarsely roughened, plants 1 to 10 cm long, branches numerous and open, lower submersed leaf-lobe about as large as the upper lobe, glochidia unbranched. - *Azolla filiculoides*
- glochidia with 1 or 2 apical septa, leaves nearly smooth, sub-orbicular, about 0.5 mm long, megaspores finely roughened, plants usually less than 3 cm long, few branches and crowded, lower submersed leaf-lobe glabrous, larger and paler than the upper lobe, glochidia may be branched. - *Azolla caroliniana*

Key to the Callitriche of British Columbia

Mature fruits are generally needed for positive identification of *Callitriche.*

- fruit encircled by a conspicuous wing-like margin, leaf bases joined by winged ridges. - *Callitriche stagnalis*
- fruit not winged or with only a narrow wing at the tip, leaf bases various.
- leaves all linear, 1-nerved, light green, leaf bases not joined by a wing, floral bracts absent, common species. - *Callitriche hermaphroditica*
- leaves various, upper often ovate and 3-nerved, bases joined by a wing-like ridge, floral bracts present.
- carpel face markings in regular vertical lines, fruit slightly wing-margined at the top and longer than broad. - *Callitriche verna*
- carpel face markings scattered, fruit not winged and as long as broad.
- fruits widest above the middle (obovate) leaves bidentate, midvein barely thickened at the end, emergent leaves may be over 5 mm wide, stems long. - *Callitriche heterophylla*
- fruits round or oblong, midvein thickened at the tip and protruding, leaves never over 5 mm wide, plants short and slender. - *Callitriche anceps*

Key to the Ceratophyllum of British Columbia

- leaf segments sub-capillary, mostly entire, delicate and light green, in deeper water and not surfacing, achene with 3 to 5 lateral spines on each side, not a 'weedy' species. - *Ceratophyllum echinatum*

- leaf segments capillary to linear and flattened, serrate to coarsely toothed, plant usually coarse and robust, dark green to almost black, usually surfacing, achene without lateral spines, 2 basal spines and 1 terminal spine only, a very 'weedy' species in eutrophic waters. - *Ceratophyllum demersum*

Key to the Chara of British Columbia

- stem not corticate, stipulodes in 1 tier, and alternating with the branchlets, branchlets tipped with a corona, no spine cells. - *Chara braunii*
- stems, and at least basal branchlet segment, corticate.
- spine cells in fascicles, axial cortex haplostichous, 1 corticate, stipulodes in two tiers, spiny looking, found in saline or brackish waters. - *Chara canescens*
- spine cells solitary or geminate, axial cortex diplostichous or triplostichous, two or 3 corticate.
- axial cortex diplostichous, stipulodes in 2 tiers, spine cells absent or small, rarely geminate, quite variable and widespread in hard-water lakes, smooth but marl encrusted. - *Chara vulgaris*
- axial cortex triplostichous, stipulodes usually in 2 tiers but may be in 1 tier, obscure or absent, spine cells usually absent or obscure but may be clustered, variable and widespread in hard-water and soft-water lakes, smooth appearance.

Chara globularis

Key to the Elodea of British Columbia

- leaves in whorls of 3.
- leaves rarely over 1.5 cm long.
- leaves 2 (1 to 4) mm wide, tapered abruptly to a blunt point, staminate flowers stalked and persistent. - *Elodea canadensis*
- leaves 1.5 (0.3 to 1.5) mm wide, tapered to a slender point, staminate flowers sessile and deciduous at anthesis. - *Elodea nuttallii*
- leaves (1.7) 2.0 to 2.6 cm long. - {*Elodea longivaginata*}
- leaves in the upper and middle part of the stem in pairs, lower ones irregular or alternate, (1.7) 2.0 to 2.6 cm long. - {*Elodea longivaginata*}

Key to the Aquatic Equisetum of British Columbia

This sub-set of all the *Equisetum* in BC share the following characteristics. Stems annual, usually with whorls of branches, cones blunt, fertile and sterile stems alike, ridges of the stem smooth or without tubercles or spicules, often cross-wrinkled, cones appear in summer.

- central cavity well over 1/2 the diameter of the stem, teeth may be deciduous, 10 to 40 ridges, teeth persistent, black but not hyaline margined, stomates in 1 row and not sunken. - *Equisetum fluviatile*
- central cavity less than 1/3 the diameter of the stem, teeth not deciduous, black and hyaline margined, 5 to 10 ridges. - *Equisetum palustre*

Key to the Gratiola of British Columbia

- there are 2 sepaloid bracts at the apex of the pedicel and thus apparently 7 sepals. - *Gratiola neglecta*
- pedicels without bracts and sepals thus evidently only 5. - *Gratiola ebracteata*

Key to the Hydrocotyle of British Columbia

- leaves orbicular to reniform, crenate or shallowly lobed, centrally peltate. - *Hydrocotyle verticillata*
- leaves 5 to 6 lobed nearly to the centre, not peltate. - *Hydrocotyle ranunculoides*

Key to the Isoetes of British Columbia

A compound or dissecting microscope is required to positively identify species in *Isoetes*. Magnification over 10x is needed to examine the megaspore surface.

- corms 3-lobed, plants primarily terrestrial, grow in grassy, ephemeral pools which are wet in winter.
- corms 2-lobed, plants primarily of lake bottoms but sometimes exposed in summer due to water level fluctuations.
- megaspores spiny.
- megaspores vary greatly in size and are often aborted, spines blunt and dense. (this is probably a hybrid of *Isoetes echinospora* and *Isoetes maritima*). - *Isoetes truncata*
- megaspores uniform in size, spines sharp, not crowded.
- plants flaccid, megaspore spines elongate with pointed tips, uniform in size, microspores smooth. - *Isoetes echinospora*

- plants rigid, megaspore spines stubby and blunt, shorter and denser along the ridges, microspores spiny. - *Isoetes maritima*
- megaspores smooth or with ridges but not spiny.
- lowermost leaves in 2 ranks, megaspores over 0.5 mm in diameter, microspores rugose and over 36μm long. - *Isoetes occidentalis*
- lowermost leaves spirally arranged, megaspores less than 0.5 mm in diameter, microspore spiny and less than 30μm long.
- base of the leaves blackened, hyaline wing margins of the leaves extending 1 to 5 cm above the sporangium, sporangia brown spotted. - *Isoetes howellii*
- base of the leaves green, hyaline wing margins of the leaves not extending more than 1 cm above the sporangium, sporangia without colour. - *Isoetes bolanderi*

Key to the Lemna *of British Columbia*

- fronds oblong to lanceolate, 6-12 mm long and connected in small groups by stalks of the same length, matted, generally submersed, colonies. *Lemna trisulca*
- fronds oval to round, less than 6 mm long, no stalks, solitary or in small attached groups, floating on the surface, usually 1 nerved, 2 to 4 mm long. - *Lemna minor*

Key to the Limnobium *of British Columbia*

- floating leaves long-petiolate, leaves flat on the top with a thin spongy layer beneath and red-brown markings on the upper surface. - *Limnobium spongia*
- floating leaves short-petiolate, leaves arched on the top with a thick spongy layer beneath and green with no coloured markings on the upper surface. - *Limnobium laevigatum*

Key to the Aquatic Myosotis *of British Columbia*

- corolla limb 2 to 5 mm wide, style shorter than the nutlets, not stoloniferous, often recumbent. - *Myosotis laxa*
- corolla limb 5 to 10 mm wide, style as long as or longer than the nutlets, stoloniferous, mostly erect and not creeping or recumbent. - *Myosotis scorpioides*

Key to the Myriophyllum *of British Columbia*

Submersed leaves are simply pinnate. The key includes both native and two introduced species, one of which is widespread in aquaria and outdoor garden pools.

- flowers in the axils of cauline submersed leaves.
- leaves in whorls of 3 to 4, or scattered, fewer than 10 leaf segments on each side of the rachis, monoecious, 4 stamens, plants often reddish, fully submersed. - *Myriophyllum farwellii*
- leaves in whorls of 4 to 6, more than 10 leaf segments on each side of the rachis, dioecious, 8 stamens, plants a pallid yellowish-green, apical portion of stem often sprawled over the surface of the water or the adjacent shore. - {*Myriophyllum aquaticum*}
- flowers in the axils of bracts on emergent, terminal spikes.
- four stamens, 4 to 6 leaves per whorl, bracts conspicuous.
- floral bracts delicate, deeply incised to serrate, spike short and delicate, widespread in sloughs of the lower Fraser Valley. - *Myriophyllum hippuroides* and *Myriophyllum pinnatum*
- floral bracts ovate and toothed, spike long, robust and inflated, introduced in several park and garden ponds of south-western British Columbia. - {*Myriophyllum heterophyllum*}
- eight stamens, 3 to 5 leaves per whorl, floral bracts various.
- floral bracts smaller than the flowers, inconspicuous, and nearly entire (the lowest few may be larger and pinnate but the upper ones are small), leaf whorls in the central portion of the stem are over 1 cm apart and not crowded, monoecious.
- no turions, rhizomatous, 10 to 16 leaf divisions less than 2 mm apart, leaves make right or obtuse angles with the stem, leaf tips 'squared', all leaf segments straight and all of nearly the same length. - *Myriophyllum spicatum*
- turions present, not rhizomatous, 6 to 12 leaf segments over 2.5 mm apart, leaves make acute angles with the stem, leaf tips 'acute', basal leaf segments curved and much longer than the apical. - *Myriophyllum sibiricum* [*Myriophyllum exalbescens*]
- floral bracts usually longer than the flowers and rarely entire, the leaf whorl spacing varies, monoecious or dioecious.
- dioecious (male and female flowers on separate small plants) found only on exposed mud banks when the water level drops in summer, female bracts and leaves entire to scarcely and irregularly divided, male bracts and leaves entire to pectinate-pinnate, submersed leaves scattered and irregular. - *Myriophyllum ussuriense*

- dioecious or monoecious, flowering plants found in water, bracts large and conspicuous, shape variable.
- floral bracts pinnate to pectinate, greenish, leaves often crowded on the stem and delicate, usually more than 10 leaf divisions. - *Myriophyllum verticillatum*
- floral bracts pectinately parted below becoming dentate in the middle and almost entire above, reddish, leaves well spaced on the stem and robust, generally.

Key to the Nitella of British Columbia

- the end segment of the branchlets is one-celled.
- heteroclemous, 2 kinds of branchlets in a whorl, 1-forked alternating with one-celled, bright green, compact, branchlets swollen. - *Nitella clavata*
- homeoclemous, all branchlets in a whorl the same, generally 1-forked.
- fertile heads small and densely compact, no mucous, generally no terminal dactyls on the sterile branchlets. - *Nitella acuminata*
- fertile heads absent or loose and not compact.
- dactyl apices not long acuminate, may be acute, blunt or apiculate, (a small variety may have 2-forked branchlets occasionally). - *Nitella flexilis*
- dactyl apices are long acuminate, gametangia on branchlets, generally not marl covered. - *Nitella acuminata*
- the end segment of the branchlets is 2-or-more-celled, the end cells are very much smaller than the penultimate cell, branchlets furcate, heads terminal, branchlets homeoclemous.
- robust plants, over 15 cm high, axes over 600 microns in diameter, dactyls may be small. - *Nitella furcata*
- smaller plants, seldom over 15 cm tall, axes 200 to 450 microns in diameter, dactyls uniform.
- lowest branchlet node fertile, dactyls 2- to 3-celled, heads may have mucous, oospore membrane granular or felt-like. - *Nitella gracilis*
- lowest branchlet node sterile, dactyls 2-celled, heads without mucous, oospore membrane reticulate. - *Nitella tenuissima*

Key to the Nuphar of British Columbia

In *Nuphar* the large, showy, yellow, perianth members are sepals; the petals are smaller than the stamens and inconspicuous.

- sepals 6 to 8 (usually 6), 2.5 to 3.5 cm long, stamens yellow. - *Nuphar variegatum*
- sepals 8 to 17 (usually 9), (3) 3.5 to 6 cm long, stamens reddish. - *Nuphar polysepalum*

Key to the Nymphaea of British Columbia

- flowers yellow, 6 to 13 cm in diameter, leaves dark green and blotched above, brownish with black dots below, plant tuberous. - *Nymphaea mexicana*
- flowers white pink or red, leaves not blotched.
- small plant with leaves up to 8 cm wide and flowers up to 5 cm wide, petals seven to 15, stigmas 6 to 9, flowers white. - *Nymphaea tetragona*
- larger plants with leaves over 8 cm wide and flowers over 5 cm wide, petals and stigmas numerous, flowers white, pink, or rose.
- flowers white or tinged with pink, 20 to 32 petals, leaves scattered on the rhizome, flowers strongly scented, leaves usually purplish beneath. - *Nymphaea odorata*
- flowers white, 12 to 24 petals, leaves crowded on the rhizome, flowers not strongly scented, leaves greenish beneath. - *Nymphaea alba*

Key to the Nymphoides of British Columbia

- flowers yellow, axillary, no cluster of roots on the petiole, leaves primarily from branching stems. - *Nymphoides peltata*
- flowers white, in clusters on the petioles, a cluster of roots on the petiole, leaves mostly basal with long slender petioles.
- leaf blades ovate, mostly under 5 cm long, seeds smooth, leaves not dark-punctate beneath. - *Nymphoides cordatum*
- leaf blades orbicular, 8 to 15 cm in diameter, seeds glandular, warty, leaves dark-punctate or pitted beneath. - *Nymphoides aquaticum*

Key to the Phyllospadix of British Columbia

- fertile stems branched, 4 to 12 dm long, spathes usually paired at the nodes. - *Phyllospadix torreyi*
- fertile stems unbranched, 0.5 to 4 dm long, spathes usually solitary at the nodes.

- leaves with 3, rarely 5, veins, margins entire. - *Phyllospadix scouleri*
- leaves with 5 or 7 veins, margins toothed towards the apex. - *Phyllospadix serrulatus*

Key to the Aquatic Polygonum of British Columbia

All our BC aquatic *Polygonum* share the following characteristics. Stipules red or brown, cylindrical or funnel-like, sometimes bristly, usually perennial and rhizomatous, fresh water.

- flowers rose in 1 or 2 terminal inflorescences, perianth not glandular, achenes lenticular. - *Polygonum amphibium*
- flowers whitish, pinkish or greenish-white in 2-to-many contracted panicles, perianth may be glandular, achenes mostly triquetrous.
- perianths glandular-punctate, glands sessile, stamens 6 or 8, achenes brown, glandular and dull, perianth usually 4-lobed, stamens usually 6. - *Polygonum hydropiper*
- perianth not glandular-punctate, achenes dark brown or black, stamens 8.
- inflorescence ending in spicate, slender, interrupted racemes mostly over three cm long, nerves of the perianth segments not branched and recurved, stamens 8, perianth 5-lobed. - *Polygonum hydropiperoides*
- inflorescence of several short, thick, continuous racemes, rarely over three cm long, nerves of the perianth segments branched and recurved, perianth 4 to 5 lobed. - *Polygonum lapathifolium*

Key to the Potamogeton of British Columbia

- stipules forming a sheath around the stem partly below the base of the leaf blade and partly above, the leaf attached 1 cm or more above the node.
- leaves 3 to 4 mm wide, flat, rough on the edges, stiff, with a broad mid-vein and over 20 lateral veins.
- leaves less than 3 mm wide, cylindrical, smooth, soft, with only one inconspicuous vein.
- leaves acute and sharp pointed at the apex, fruits 2.5 to 4 mm long with a short beak, the surface of fruits without radial striation (at 10x). - *Potamogeton pectinatus*
- leaves obtuse and blunt at the apex, fruits 2 to 3 mm long, beakless, surface of the fruits with radial striation (at 10x).

- primary stems 0.5 to 1 mm in diameter, all leaves filiform, 0.2 to 0.5 mm wide with tight sheaths, spikes with 2 to 5 whorls of flowers, fruits 2 to 2.5 mm long. - *Potamogeton filiformis*
- primary stems 1 to 3 mm in diameter, lower main stem leaves with blades 1 to 2 mm wide and with loose inflated sheaths, upper branch leaves filiform, spikes with 5 to 12 whorls of flowers, fruits 2.5 to 2.8 mm long. *Potamogeton vaginatus*
- stipules forming a sheath around the stem only above the base of the leaf blade, the leaf attached at the node.
- submersed leaves linear, ribbon-like or cylindrical, less than 5 mm wide, parallel margins.
- submersed leaves cylindrical, terete.
- floating leaves 2.5 to 6 cm wide, usually cordate, large ones with (18) 21 to 35 nerves, submersed leaves arise from the main stem, mature fruit 3.7 to 4.5 mm long including the beak, obscurely keeled, interlacunar bundles in several circles throughout the aerenchyma. - *Potamogeton natans*
- floating leaves 1 to 3 cm wide, usually rounded or cuneate at the base, largest ones with 11 to 19 nerves, submersed leaves arise from the branches of the main stem, mature fruit, including the beak, 3 to 3.4 mm long, interlacunar bundles develop only in the outer circle of the aerenchyma. - *Potamogeton oakesianus*
- submersed leaves flat.
- submersed leaves with 9 to 35 veins. - *Potamogeton zosteriformis*
- submersed leaves with 1 to 7 veins.
- stipules strongly fibrous, becoming whitish, especially on the turions, base of turions strongly ribbed.
- leaves obtuse or rounded and slightly mucronate, not conspicuously 2 ranked,, blades thin, 1.5 to 3.5 mm wide, 5 to 7 veins, a narrow cellular-reticulate band along the midrib, turions fan-shaped, peduncle flattened. - *Potamogeton friesii*
- leaves gradually tapered into a sharp bristle tip, conspicuously 2 ranked, blades firm, 0.5 to 2.5 mm wide, convolute with 3 (5) veins, no cellular-reticulate band along the midrib, turions slender, peduncles terete. - *Potamogeton strictifolius*
- stipules delicate, not fibrous, greenish or brownish, base of the turions smooth.

- leaves (2) 3 to 4 mm wide, rounded at the apex, fruits (3) 3.5 to 4 mm long. - *Potamogeton obtusifolius*
- leaves 0.3 to 3 mm wide, acute to obtuse or mucronate, fruits 1.8 to 2.8 mm long
- fruits with a distinct dorsal keel, veins on the stipules evident as ridges running the full length of the stipule, glands at the base of the stipules either lacking or poorly developed. - *Potamogeton foliosus*
- fruits without a dorsal keel, veins on the stipules obscure and faint, glands at the base of the stipules usually well developed.
- stipules connate, fused along the stem, at least when young, mature fruits broadest above the middle, plants sparsely branched. - *Potamogeton pusillus*
- stipules convolute, mature fruits broadest below the middle, plants well branched. - *Potamogeton berchtoldii*
- submersed leaves lanceolate or ovate, more than 5 mm wide.
- leaf margins serrate, beak of the fruit as long as the fruit body or longer. - *Potamogeton crispus*
- leaf margins entire, beak of the fruit much shorter than the fruit body.
- submersed leaves ribbon-like with parallel side, 5 to 10 mm wide, limp and flaccid, median band of lacunae several cells wide at least 1/4 the width of the blade, stems compressed. - *Potamogeton epihydrus*
- submersed leaves lanceolate to ovate with sides not parallel, either no median cellular-reticulate band or the band less than 1/4 the width of the blade, stems terete.
- submersed leaves sessile, cordate or rounded at the base and clasping the stem.
- leaves ovate-oblong, mostly 10 to 20 cm long, with a cucullate apex, fruits more than 4 mm long, stems with many interlacunar vascular bundles. - *Potamogeton praelongus*
- leaves rounded, ovate or elongate-ovate, 1 to 10 cm long, no cucullate apex, fruits under 3.5 mm long, no interlacunar vascular bundles.
- stipules coarse disintegrating into persistent whitish fibres, peduncles clavate, 1.5 to 25 cm long, fruits with a cavity in the endocarp loop. - *Potamogeton richardsonii*

- stipules delicate, lacking on old specimens, peduncles not clavate, 1 to 9 cm long, fruits without a cavity in the endocarp loop. - *Potamogeton perfoliatus*
- submersed leaves petiolate or sessile, not clasping the stems.
- submersed leaves 2 to 5 cm wide, tapering into distinct petioles, stipules over 3 cm long.
- floating leaves cordate, submersed leaves folded and strongly falcate, more than 20 veins, interlacunar bundles well developed throughout. - *Potamogeton amplifolius*
- 20. floating leaves rounded or cuneate at the base, submersed leaves flat and not falcate, with fewer than 20 veins, interlacunar bundles developed in only one circle or absent.
- petioles of submersed leaves shorter than 2 cm, interlacunar bundles well developed forming one circle, endodermis of U-cells. - *Potamogeton illinoensis*
- petioles of submersed leaves longer than 2 cm, interlacunar bundles absent, endodermis of O-cells. - *Potamogeton natans*
- submersed leaves mostly less than 2 cm wide, sessile, or with petioles under 0.5 cm long, stipules under 3 cm long.
- plants with a reddish tinge, usually not branched, floating leaves, if present, not markedly different from the submersed leaves, no interlacunar bundles, endodermis of O-cells. - *Potamogeton alpinus*
- plants greenish, freely branched, floating leaves markedly different from the submersed leaves, interlacunar bundles well developed in one circle, endodermis of U-cells. - *Potamogeton gramineus*

Key to the Aquatic Ranunculus *of British Columbia*

- flowers white, submersed leaves dissected into ultimately filiform segments.
- two to 7 glabrous, beakless achenes, plants and receptacle glabrous, pedicels in the axils of ternately lobed floating leaves, leaves 2 to 3 times divided into 8 to 12 segments. - *Ranunculus lobbii*
- ten to 80 glabrous or hirsute achenes, plants and receptacles hirsute.
- thirty to 80 achenes, flowers may have yellow bases, leaf blades sessile on the stipular base. - *Ranunculus circinatus*

- ten to 25 achenes, flowers all white, leaf blades petiolate. - *Ranunculus aquatilis*
- flowers yellow, submersed leaves, if present, simple to ternately divided.
- leaves simple and entire to serrate, may be shallowly lobed, achenes glabrous.
- achenes 50 to 200 in a columnar head, longitudinally ribbed. - *Ranunculus cymbalaria*
- achenes 5 to 60 in a globular head, not longitudinally ribbed, stems decumbent to prostrate, rooting at the nodes, nectary scales usually broader than long.
- leaves all, or in part, lobed, parted or ternately dissected to filiform segments, achenes pubescent.
- annuals, erect, not nodally rooting, achene beaks inconspicuous. - *Ranunculus sceleratus*
- perennials, floating or reclining, nodally rooting, achene beaks conspicuous.
- leaves deeply lobed or parted, achenes not corky margined.
- leaves deeply 3 parted with narrow lobes distally acute, achene beaks one quarter the length of the achenes, receptacles hirsute, 5 sepals and petals. - *Ranunculus gmelinii*
- leaves 3 lobed, distally rounded, achene beaks 1/10 the length of the achenes, receptacles glabrous, 3 sepals and petals. - *Ranunculus hyperboreus*
- leaves, at least the submersed ones, 3 to 5 times ternately dissected into filiform segments less than 2 mm wide, achenes corky margined. - *Ranunculus flabellaris*

Key to the Sagittaria of British Columbia

In *Sagittaria* the usual non-achene key characters of bract length and shape and pedicel lengths have been found to be inconsistent and sufficiently variable even within one population as to be virtually useless as key characters. Many flowering, but non-fruiting, collections are difficult to identify.

- leaves basally sagittate or hastate, stamens 7 to 25, achenes densely packed on the receptacle, plants emergent in shallow water. - *Sagittaria*
- monoecious, mature achenes 2.0 to 2.5 mm long with a beak less than 0.5 mm long pointing forward from the tip of the achene. - *Sagittaria cuneata*

- monoecious or dioecious, mature achenes (2.5) 3.0 to 3.5 (4.0) mm long with a beak about 1 mm long at right angles to the body of the achene. - *Sagittaria latifolia*
- leaves gradually tapered or cordate at the base; broadly ribbon-like, submersed in deep water or floating on the surface, ribbon-like or tapered at the base and oval, flaccid, plants generally sterile. - *Sagittaria*

Key to {Salvinia}

There is still considerable taxonomic confusion in *Salvinia* and since these plants are only expected as sporadic, non-persistent escapes from aquaria and garden ponds, there is little benefit in trying to identify individual species. The genus is readily recognizable but individual species are often only distinguished on technical and difficult characters. There is hybridization occurring and some 'species' are sterile hybrids and serious tropical weeds.

- the four hairs on the tips of each leaf papilla are joined at the distal end. - {*Salvinia auriculata*} complex [*Salvinia molesta*]
- the hairs on the tips of the leaf papillae are free. - other {*Salvinia*} species

The *Salvinia auriculata* complex includes, in part, such species as {*Salvinia herzogii*}, {*Salvinia biloba*}, and {*Salvinia auriculata*}

The other species include, in part, such species as {*Salvinia natans*}, {*Salvinia sprucei*}, {*Salvinia cucullata*}, {*Salvinia rotundifolia*}, {*Salvinia minima*} and {*Salvinia oblongifolia*}.

Key to the Sparganium of British Columbia

- two (1) stigmas over 2 mm long, the achenes truncate-pyriform and narrowed abruptly to the beak, inflorescence usually branched, staminate heads above the pistillate heads, large robust plants. - *Sparganium eurycarpum*
- one stigma, achenes fusiform and tapering to the beak, inflorescence usually simple, staminate heads below the pistillate heads.
- achene beaks less than 1.5 mm long or beakless, staminate head solitary.
- achene beaks less than 0.5 mm long to absent, staminate head closely adjacent to the upper pistillate head, basal leaves (0.5) 1 to 3 (5) mm wide, opaque, often yellow. - *Sparganium hyperboreum*

- achene beaks 0.5 to 1.5 mm long, staminate head distant from the upper pistillate head, basal leaves (1.5) 2 to 6 (10) mm wide, translucent, usually dark green. - *Sparganium natans*
- achene beaks 1.5 to 5 mm long, there may be more than 1 staminate head.
- achene beaks conspicuously curved. - *Sparganium fluctuans*
- achene beaks straight or slightly curved.
- staminate head usually solitary, closely adjacent to the upper pistillate head. - *Sparganium glomeratum*
- staminate heads (1) 3 to 8, distant from the upper pistillate head.
- pistillate heads (1) 2 to 3 (4), usually crowded and appearing as one elongated head terminating the inflorescence, achene beaks 1.5 to 2.0 (2.2) mm long. - *Sparganium angustifolium*
- pistillate heads (3) 4 to 7 (10), distant and distinct, achene beaks 2 to 4.5 (6) mm long. - *Sparganium emersum*

Key to the Utricularia of British Columbia

- leaves divided into ultimately terete or threadlike segments.
- leaves divided into fewer than 5 final threadlike segments, leaf margins glabrous, bladders scarce on a small delicate plant usually floating at the surface or entangled in other rooted plants. - *Utricularia gibba*
- leaves 'pinnatifid', more than 20 terete final segments, hairy leaf margins, many bladders on the ordinary leaves, a robust plant to several meters long, usually lying on the sediment surface. - *Utricularia vulgaris*
- leaves divided di- or tri-chotomously into ultimately flattened segments.
- ordinary leaves generally with a few bladders, leaf margins glabrous, the terminal leaf segments are acuminate. - *Utricularia minor*
- ordinary leaves rarely, if ever, with bladders which are found on separate subterranean branches, leaf margins hairy, the ultimate leaf segment are awned. - *Utricularia intermedia*

Key to the Aquatic Veronica of British Columbia

- leaves all short-petiolate. - *Veronica beccabunga*
- leaves, at least those on the middle and upper portions of the flowering shoots, sessile.

- capsule flattened, notched, wider than high, 5 to 9 seeds per locule, 1 to 2 mm long, leaves 4 to 20 times as long as wide. - *Veronica scutellata*
- capsule turgid, scarcely notched, barely wider than high, many seeds less than 0.5 mm long, leaves 1.5 to 5 times as long as wide.
- leaves 1.5 to 3 times as long as wide, fruiting pedicels ascending, capsules higher than wide if not equal, flowers blue or violet. - *Veronica anagallis-aquatica*
- leaves 2.5 to 5 times as long as wide, fruiting pedicels spreading, capsules wider than high, flowers white to pink or pale blue. - *Veronica catenata*

Key to the Wolffia of British Columbia

- plants floating just below the surface, sub-globose, upper surface rounded, green, not puncticulate, 0.5-1.0 mm long. - *Wolffia columbiana*
- plants floating on the surface, ellipsoidal, upper surface flattish, white- or brown-punctilate, 0.5-1.2 mm long and about 1/2 as wide. - *Wolffia borealis*

Key to the Zostera of British Columbia

- leaves 3-veined, 1 to 1.5 mm wide, sheaths split to the base, rare and probably introduced from Asia, known only from Boundary Bay and Tsawwassen. - *Zostera japonica*
- leaves 5-veined or more, 1.5 to 12 mm wide, sheaths on sterile shoots are closed at the base, widespread native species. - *Zostera marina*

Distribution, Selected Common Names and Habitat within BC

Not all of the plants in these keys are native or naturalized in British Columbia. Some are garden or aquarium plants that are introduced from time to time but do not persist, or are in cultivation but do not survive long when they escape. Others are major invasive weeds in other parts of the world and, although they have not yet been found in British Columbia, they are found in neighbouring areas and are expected to be found in British Columbia eventually. People deliberately, and inadvertently, introduce many aquatic plants, some of which do eventually become permanently established.

The following brief notes give an indication of the habitats of native and introduced aquatic and wetland plants in British Columbia

and their general distribution range. A few selected common names are also given with the preferred British Columbia common name listed first and other common names following in alphabetical order. There are literally dozens of local common names, in many languages, for some of the more widespread species of aquatic weeds. A more complete list of these common names can be found in Aquatic Plants of British Columbia: Common Names, Selected References, Synonymy and Classification by Life-Forms and Habitat. Occasionally other pertinent or interesting notes are included.

Alisma: The species in this genus are commonly known as Mud plantains, Plantains and Water plantains

Alisma Gramineum J. G. Gmel: Narrow leaved water plantain, Grass leaved water plantain Shallow water of lakes, rivers, marshes and tidal flats in the lowland, steppe and montane zones; uncommon in southern BC; circumpolar.

Alisma Plantago-aquatica L: American water plantain, Arrowhead, Broadleaf water plantain, Common water plantain, Heart shaped water plantain, Mud plantain Shallow water of ponds, marshes and ditches in the lowland, steppe and montane zones; common in southern BC south of 52 degrees N; rare northward.

Azolla: The species in this genus are commonly known as Fairy mosses, Mosquito ferns, Water velvets and Water ferns

Azolla Caroliniana Willd: Carolina water fern Surface of water in sloughs and ditches in the lowland zone; rare in south-western BC, primarily in the lower Fraser Valley. Introduced and available in garden shops, aquatic plant nurseries and the aquarium trade.

Azolla Filiculoides Lam: Large mosquito fern, Duckweed fern, Pacific azolla Surface of water in sloughs and ditches in the lowland zone; rare in south-western BC, primarily in the lower Fraser Valley. A European introduction with large populations where found. Commonly available in garden shops, aquatic plant nurseries and the aquarium trade.

Azolla Mexicana K. B. Presl: Mexican mosquito fern Surface of the water in sloughs and pools in the montane zone; rare in south-central BC, known from the Salmon Arm and Sicamous area.

Brasenia Schreberi Gmel: Watershield, Dollar bonnet, Frog leaf, Junsai, Little water lily, Purple bonnet, Purple wendock, Schreber watershield, Water target Ponds, lakes and slow-moving streams in the lowland, steppe and montane zones; common in south-western BC, less frequent in south-central BC.

{Cabomba Caroliniana Gray}: Fanwort, Carolina water shield, Fish grass, Parrot feather, Washington grass A commonly introduced aquarium and garden pond plant, sometimes dumped but not known to overwinter outdoors in BC, established in Oregon and the lower Columbia River of Washington.

Calla Palustris L: Wild calla, Bog arum, Calla lily, Water arum

Shallow water of swamps, marshes and lakes in the montane zone; frequent in BC north of 52 degrees N and east of the Coast Mountains, rare in south-central BC; circumpolar.

***Callitriche*:** The species of this genus are commonly known as Starworts, Water chickweeds and Water starworts.

***Callitriche Anceps* Fern:** Two edged water starwort Shallow water of ponds and lakes in the lowland and montane zones; rare along the coast in BC.

***Callitriche hermaphroditica* L:** Northern water starwort, Autumnal starwort Shallow water of sow streams, ditches and sloughs in the lowland, steppe and montane zones; common in south-western BC, less frequent in south-central and south-eastern BC; circumboreal.

***Callitriche Heterophylla* Pursh:** Diverse leaved water starwort, Large water starwort Shallow water of slow streams, ditches, lakes and ponds in the lowland zone; frequent on southern Vancouver Island, rare in south-eastern BC.

Callitriche Stagnalis Scop: Pond water starwort, Common starwort, Common water starwort Shallow water of slow streams, ditches and ponds; infrequent in southern BC; introduced from Europe.

Callitriche Verna L: Spring water starwort, Common water starwort, Vernal water starwort Shallow water of slow streams, ditches and shorelines in the lowland, steppe and montane zones; frequent in southern BC; circumboreal.

Caltha Natans Pallas: Floating marsh marigold Shallow water of ponds and lakes in the montane zone; locally frequent in north-eastern BC; amphiberingian.

Caltha Palustris L: Yellow marsh marigold, Cowslip Wet sites, bogs and shallow water in the lowland zone; rare along the coast.

Ceratophyllum: The species in this genus are commonly known as Coontails, Foxtails, Hornworts and Water mosses.

***Ceratophyllum Demersum* L:** Common coontail, Common hornwort, Grey pimpled hornwort, Morass weed Ponds, lakes and slow streams in the lowland, steppe and montane zones; common in southern BC, less frequent northwards.

***Ceratophyllum Echinatum* Gray:** Spring hornwort, Prickly coontail, Spiny hornwort Lakes and sloughs in the lowland and montane zones; locally frequent on southern Vancouver Island and the Gulf Islands, rare in the lower Fraser Valley and along the Alaska Highway.

***Chara*:** The species of this genus are commonly known as Chara, Musk grasses, Skunk grasses and Stoneworts.

***Chara Braunii* Gm:** Little data is available on the abundance and distribution of charophytes in BC.

***Chara Canescens* Desv. and Lois:** Little data is available on the abundance and distribution of charophytes in BC.

***Chara Globularis* Thuill:** Little data is available on the abundance and distribution of charophytes in BC.

Chara Vulgaris L: Little data is available on the abundance and distribution of charophytes in BC.

Crassula Aquatica (L.) Schoenl: Pygmyweed Vernal pools and mud flats in the lowland zone; rare, scattered throughout southern BC; circumpolar. [Tillaea aquatica L.]

Dulichium Arundinaceum (L.) Britt: Three way sedge, Dwarf bamboo, Pond sedge Wet meadows, shallow water of lakes and stream sides in the lowland, steppe and montane zones; frequent in south-western BC, rare in south-central BC.

{Egeria Densa Planch.}: Brazilian waterweed, Anacharis, Argentine water weed, Brazilian elodea, Dense water weed, Giant elodea, Leafy elodea, Oxygen weed, South american elodea Lakes, ponds and streams in the lowland zone; known from only a few sites in Victoria and Vancouver; introduced from South America. Naturalized and prolific where present and of concern as a weed displacing native species. Readily available in garden centres and aquarium shops. Well established in over a dozen lakes in Washington State.

{Eichhornia Crassipes (Mart.) Solms}: Water hyacinth, Florida devil An often introduced garden pond species, but not yet known to overwinter outdoors in BC. Readily available in garden centres and aquarium shops. (None of the authors specimens, which grow and multiply prolifically in the greenhouse and outdoors in summer, has ever survived outdoors over the winter in Victoria).

Elatine Rubella Rydb: Three flowered waterwort Ditches, mud flats, shallow ponds and lake shores in the lowland zone; rare in south-western BC.

Elodea: The species of this genus are commonly known as Ditch mosses, and Waterweeds.

Elodea Canadensis Rich: Canadian waterweed, American elodea, Canadian pondweed, Common elodea, Water thyme Lakes, ponds, streams and ditches in the lowland, steppe and montane zones; common in south-western BC, rare northward.

Elodea Longivaginata St. John: Long leaved waterweed This species has not been confirmed in BC yet but is found east of the Rockies in Alberta, Montana and Wyoming. It is included in the keys in case a specimen is found in south-eastern BC or is introduced with garden plants. This species should be able to establish and overwinter successfully if introduced

Elodea Nuttallii (Planch.) St. John: Nuttall's waterweed, Slender waterweed, Western waterweed Lakes, ponds and streams in the lowland, steppe and montane zones; rare in southern BC.

Equisetum: The species of this genus are commonly called horsetails or scouring rushes.

Equisetum Fluviatile L: Swamp horsetail, Water horsetail Shallow water at lake margins in the lowland, steppe and montane zones; frequent throughout BC; circumpolar. This species often forms very extensive colonies with a distinctive horizontally-banded appearance en masse.

Equisetum Palustre L: Marsh horsetail Shallow water of marshes and swamps, stream banks and forests in the lowland, steppe and montane zones; frequent in southern BC, infrequent elsewhere; circumpolar. Commonly found marginally in subalpine lakes throughout BC.

Gratiola: The species of this genus are commonly known as hedge hyssops.

Gratiola Ebracteata Bentham: Bractless hedge hyssop Wet sites and shallow marginal water in the lowland zone; frequent in south-western BC.

Gratiola neglecta Torr: Common american hedge hyssop, Clammy hedge hyssop, Obscure hedge hyssop Wet sites and shallow marginal water in the lowland, steppe and lower montane zones; rare in southern BC.

Heteranthera Dubia (Jacq.) Macmill: Water star grass, Mud plantain, Water star weed Ponds, lakes and slow-moving streams in the steppe and montane zones; infrequent in south-western and south-central BC.

Hippuris Vulgaris L: Common mare's tail, Four leaved mare's tail, Water mare's tail Shorelines, lakes, ponds and quiet open water in the lowland, steppe and montane zones; common throughout BC, less frequent along the coast; circumpolar.

{Hydrilla Verticillata (L.) Royle}: Hydrilla, Florida elodea, Oxygen grass, Oxygen weed, Water thyme Lakes, ponds streams and other permanent water. This plant has not been found in BC yet, but it is established in Washington State. It is expected to overwinter successfully and become a weed once it is introduced.

Hydrocotyle Ranunculoides L. f: Floating water pennywort Ponds, shallow water of marshes and wet sites in the lowland zone; rare on south-eastern Vancouver Island.

Hydrocotyle Verticillata Thunb: Whorled water pennywort Streams, shallow water of marshes and wet sites in the lowland zone; rare, known from the lower Fraser valley.

***Isoetes*:** The species of this genus are commonly known as Quillworts, Bracksen grasses, Merllyn's grasses or Octopus plants.

Isoetes Bolanderi Engelm: Bolander's quillwort Shallow water of lakes and muddy sites in the subalpine zone; rare in south-eastern BC, known only from Akamina Pass.

Isoetes Echinospora Dur: Bristle like quillwort Lakes in the lowland to subalpine zones; infrequent throughout BC; circumpolar.

Isoetes Howellii Engelm: Howell's quillwort Shallow water of lakes (exposed in summer) in the steppe and montane zones; rare in south-central BC, known from Kamloops, Mara and Shuswap Lakes.

Isoetes Maritima Underw: Coastal quillwort Shallow water of lakes and other shallow water habitats in the lowland and montane zones; frequent in coastal BC, rare in south-central BC; amphiberingian.

Isoetes Nuttallii A. Br: Nuttall's quillwort Vernal pools and ephemeral winter seepages in the lowland zone; rare on south-eastern Vancouver Island.

Isoetes Occidentalis Henderson: Western quillwort Lakes in the lowland, steppe and montane zones; frequent in southern BC.

Isoetes Truncata (A. A. Eaton) Clute: Slashed quillwort Shallow water of lakes (exposed in summer) in the lowland zone; infrequent on south-eastern Vancouver Island (may be a hybrid between Isoetes echinospora and Isoetes maritima).

Juncus Supiniformis Engelm: Spreading rush Shallow water, wet muck, lake shores and open bogs in the lowland and montane zones; common in BC west of the Coast-Cascade Mountains.

Lagarosiphon Major Ridley: Lagarosiphon, African elodea, Oxygen weed This 'weedy' species is distributed as an aquarium plant but it has not yet been found out of cultivation in BC.

Lemna: The species of this genus are commonly known as Duckmeats, Duckweeds and water lentils.

Lemna Minor L: Common duckweed, Lesser duckweed, Small duckweed, Water lentil Ponds, lakes and slow moving streams in the lowland, steppe and montane zones; common throughout BC south of 55 degrees N, less frequent northward, absent in north-western BC and the Queen Charlotte Islands; circumpolar.

Lemna Trisulca L: Star duckweed, Chain of stars, Forked duckweed, Ivy leaved duckweed Ponds, lakes and slow moving streams in the lowland, steppe and montane zones; frequent throughout BC, absent in the Queen Charlotte Islands; circumpolar.

Lilaea Scilloides (Poir.) Haum: Flowering quillwort Mud flats, ponds, shallow water of lakes and wet sites in the lowland zone; rare in coastal BC, bipolar disjunct.

Lilaeopsis Occidentalis Coult. and Rose:Western lilaeopsis Shallow water of marshes, lakes and tidal shores in the lowland zone; infrequent along the coast.

{Limnophila Sessiliflora Blume}: Limnophila This species is distributed as an aquarium plant but it has not yet been reported out of cultivation in BC.

{Limnobium Laevigatum} (Humb. and Bonpl.) Heine: Amazonian frogbit This species is distributed as an aquarium plant but it has not yet been reported out of cultivation in BC, [*Limnobium stoloniferum*].

{Limnobium Spongia} (Bosc) Steud: American frogbit This 'weedy' species is widely distributed as an aquarium plant but it has not yet been reported out of cultivation in BC.

Limosella Aquatica L: Water mudwort, Limosella, Mudwort Wet sites and shallow water in the lowland, steppe and montane zones; infrequent throughout BC south of 57 degrees N, absent from the Queen Charlotte Islands; circumpolar.

Lobelia Dortmanna L: Water lobelia Shallow water with a sandy, gravelly or otherwise firm bottom in lakes and ponds in the lowland zone; locally common in south-western BC.

Ludwigia Palustris (L.) Ell: Water purslane, American seedbox, Creeping primrose, False loosestrife, Low seedbox, Marsh ludwigia, Marsh purslane, Marsh seedbox, Phthisis weed Wet sites, shallow water of lakes and ponds in the lowland zone; rare on southern Vancouver Island and the lower Fraser Valley.

Lysimachia thyrsiflora L: Tufted loosestrife, Water loosestrife Shallow water of marshes, lakes and ponds in the lowland and montane zones; frequent in BC.

Marsilea Vestita Hook. and Grev: Hairy water clover, Clover fern, Hairy pepperwort, Water shamrock, Western water clover Inundated lake margins (exposed in the summer) in the steppe and montane zones; rare in south-central BC, known from around Vernon.

Megalodonta Beckii (Torr. ex Spreng.) Greene: Water marigold, Bur marigold Lake shores and shallow open water; rare known from south-eastern BC and Vancouver Island. [*Bidens beckii* (Torr. ex Spreng.)].

Menyanthes Trifoliata L: Buckbean, Bog bean, Marsh trefoil Bogs, shallow water of ponds and lakes and lake inlet and outlet fans in the lowland, montane and steppe zones; common throughout BC; circumpolar.

Montia Fontana L: Water chickweed, Blinks Wet meadows or shallow water in the lowland and montane zones; rare in north-western, south-western and south-central BC; circumpolar. *Myriophyllum* The species of this genus are commonly known as Frills, Milfoils, Parrot feathers, Quills, and Water milfoils.

{Myriophyllum Aquaticum (Vell.) Verd.}: Brazilian water milfoil, Golden myriophyllum, Parrot feather, Water feather Shallow water of garden ponds and drainage ditches in the lowland zone; infrequent on southern Vancouver Island and the lower Fraser Valley; introduced from South America. *Myriophyllum aquaticum* is established in several lakes and rivers in Washington and Oregon, 2 lakes are in northwest Washington. [*Myriophyllum brasiliense* Cambess. in Hill *et al.*]

Myriophyllum Farwellii Morong: Farwell's water milfoil Lakes and sloughs in the lowland and lower montane zones; infrequent, scattered throughout southern BC.

{Myriophyllum Heterophyllum Michx.}: Various leaved water milfoil, Broad leaf water milfoil, Foxtail, Mare's tail, Mermaid weed Lakes and ponds in the lowland zone; rare, found in Beaver Lake in Stanley Park (now extirpated) and several ponds in Queen Elizabeth Park in Vancouver; introduced from eastern North America.

Myriophyllum Hippuroides Nutt: Western water milfoil, Brown myriophyllum, Red water milfoil, Variable water milfoil Lakes, ponds, sloughs and ditches in the lowland zone; infrequent, known from the lower Fraser Valley.

Myriophyllum Pinnatum (Walt.) B. S. P: Green parrot's feather, Cut leaf water milfoil, Eastern water milfoil, Variable water milfoil Lakes, ponds, sloughs and ditches in the lowland zone; infrequent, known from the lower Fraser Valley. [*Myriophyllum scabratum* Michx.]

Myriophyllum Quitense H. B. K: Waterwort water milfoil, Andean water milfoil, Coarse water milfoil Lakes and rivers in the lowland zone; rare on Vancouver Island.

Myriophyllum Sibiricum Kom: Siberian water milfoil, American water milfoil, Northern water milfoil, Spiked water milfoil, Western spiked water milfoil Lakes, ponds, rivers and other water bodies in the lowland and montane zones; frequent throughout BC; circumpolar. [*Myriophyllum exalbescens* Fern.]

Myriophyllum Spicatum L: Eurasian water milfoil, Fox tail, Spiked water milfoil Lakes, ponds, sloughs, irrigation ditches and other water bodies in the lowland, steppe and montane zones; infrequent, scattered throughout southern BC; introduced from Eurasia.

Myriophyllum Ussuriense (Regel) Maxim: Ussurian water milfoil Marginal habitats of lakes and rivers subject to seasonal water level fluctuations, and exposure in late summer, in the lowland and lower montane zones; rare and scattered on Vancouver Island, the lower Fraser Valley and south-central BC; amphiberingian.

Myriophyllum Verticillatum L: Verticillate water milfoil, Green milfoil, Whorled water milfoil Lakes and sloughs of the lowland to montane zones; frequent throughout BC; circumpolar.

Najas Flexilis (Willd.) R. and S: Wavy water nymph, Common naiad, Northern naiad, Slender naiad, Slender water nymph Ponds and marshes in the lowland, steppe and montane zones; common in southern BC; circumpolar.

Nasturtium Officinale R. Br. in W. Ait: Common water cress, True water cress Streams, ditches, swamp margins and shallow ponds in the lowland and montane zones; frequent in southern BC, rare northward; introduction from Europe. [*Rorippa nasturtium-aquaticum* (L.) Hayek]

***Nitella*:** The species of this genus are commonly known as Muskgrasses, Nitellas, Sandgrasses, Skunkgrasses, and Stoneworts.

Nitella Acuminata A. Br: Very little data is available on the abundance and distribution of charophytes in BC.

Nitella Clavata Kutz: ery little data is available on the abundance and distribution of charophytes in BC.

Nitella Flexilis (L.) Ag: Very little data is available on the abundance and distribution of charophytes in BC.

Nitella Furcata (Roxb.) Ag: Very little data is available on the abundance and distribution of charophytes in BC.

Nitella Gracilis (Sm.) Ag: Very little data is available on the abundance and distribution of charophytes in BC.

Nitella Tenuissima (Desv.) Kutz: Very little data is available on the abundance and distribution of charophytes in BC.

Nuphar: The species of this genus are commonly known as Yellow lilies, Cow Lilies, Pond lilies, Water lilies, Spatterdocks and Water collards.

Nuphar Polysepalum Engelm: Rocky mountain cow lily, Indian pond lily, Western pond or water lily Lakes, ponds and slow-moving streams in the lowland, steppe and montane zones; common in BC south of 55 degrees N and west of 120 degrees W, less common northward and eastward.

Nuphar Variegatum Engelm: Bullhead pond lily, Common yellow pond lily, Variegated pond lily Lakes, ponds and slow-moving streams in the steppe and montane zones; frequent in eastern BC.

Nymphaea: The species of this genus are commonly known as Fragrant water lilies, Pond lilies, Water nymphs, Water lilies and White water lilies

Nymphaea Alba L.: European white water lily Ponds and lakes in the lowland and montane zones; rare on south-eastern Vancouver Island and at Clearwater; introduced from Europe.

Nymphaea Leibergii Morong: Pygmy water lily, Dwarf water lily, Little white water lily, Northern water lily, Small white water lily Lakes, ponds and slow-moving streams in the montane zone; rare and scattered in eastern BC.

Nymphaea Mexicana Zuccarini: Yellow water lily, Banana water lily, Yellow lotus Ponds and lakes in the lowland zone; rare on south-eastern Vancouver Island; introduced from Mexico.

Nymphaea Odorata Ait: Fragrant water lily, Alligator bonnet, American water lily, Cow cabbage, Large white water lily, Sweet scented water lily, Toad lily, Water cabbage Lakes, ponds and slow-moving streams in the lowland, steppe and lower montane zones; rare and scattered in south-western and south-central BC; introduced from Europe.

Nymphaea Tetragona Georgi: Pygmy water lily, Dwarf water lily, Little white water lily, Northern water lily, Small white water lily Lakes, ponds and slow-moving streams in the lowland and montane zones; rare and scattered from the central coast and interior northward; circumpolar.

Nymphoides Aquatica (Gmelin) O. Kuntze: Banana plant, Fairy water lily This species is found in garden ponds but has not yet been found naturalized.

{Nymphoides Cordatum} (Ell.) Fernald: Floating heart This species is found in garden ponds but has not yet been found naturalized.

Nymphoides Peltata (Gmelin) O. Kuntze: Water fringe, Floating heart This species is found in garden ponds but has not yet been found naturalized in BC, it is established in Washington State near Spokane.

Phyllospadix: The species of this genus are commonly known as Sea grasses or Surf grasses.

Phyllospadix Scouleri Hooker: Scouler's surf grass Exposed, rocky, intertidal to subtidal shores in the lowland zone; common along the coast.

Phyllospadix Serrulatus Rupr. ex Aschers: Toothed surf grass Sheltered intertidal and subtidal shores in the lowland zone; frequent along the coast.

Phyllospadix Torreyi S. Wats: Torrey's surf grass Exposed, rocky, intertidal to subtidal shores in the lowland zone; infrequent in coastal BC.

{Pilularia Americana R. Br.}: American pillwort Shallow water of ponds in the lowland zone; rare in south-western BC; introduced from south-western North America, known from a pond in the UBC botanical gardens.

{Pistia Stratiotes L.}: Water lettuce, Shell flower, Water cabbage, Water bonnet This is a free-floating tropical species introduced in the aquarium and garden pool trade but it has not yet been found out of cultivation It is not expected to overwinter outdoors in BC. (None of the authors specimens have ever overwintered successfully. There is a long list of endemic common names, mostly in southeast Asian languages and tropical American tongues).

Polygonum: The species in this genus are commonly known as Door weeds, Knot weeds, Smart weeds and Tear thumbs.

Polygonum Amphibium L: Water smartweed, Amphibious bistort, Water smartweed, Floating knotweed, Marsh smart weed, Swamp persicaria, Swamp smart weed, Water heartsease, Water knot weed, Water persicaria Shorelines, ditches and shallow water of lakes and ponds in the lowland, steppe and montane zones; common throughout BC except for the Queen Charlotte Islands and adjacent coast.

Polygonum Hyropiper L: Marshpepper smartweed, Annual smartweed, Common smartweed, Water smartweed, Marsh pepper, Water pepper

Moist ditches, shallow marginal sites and disturbed places in the lowland, steppe and montane zones; infrequent in south-western and south-central BC; introduced from Eurasia.

Polygonum Hydropiperoides Michx: Swamp smartweed, Marshpepper smartweed, Mild water pepper, Wild water pepper Wet swampy sites, shorelines and shallow marginal water of lakes and rivers in the lowland, steppe and montane zones; rare in southern BC.

Polygonum Lapathifolium L: Willow weed, Nodding smartweed. Pale smartweed Wet swampy sites, wet meadows, shorelines and shallow marginal water of lakes and rivers in the lowland, steppe and montane zones; common in south-western BC, rare elsewhere in southern and north-eastern BC.

{Pontederia Cordata Lour.}: Pickerel weed, Pikeweed, Wampee This introduced subtropical species is often found in garden pools and is known from Glen Lake in Victoria where it has overwintered successfully for over a decade, but not spread from its original site. (The author has kept a population growing, outdoors, in a tub of water for many years; it suffered severely during the cold winter of 1995 but survived and recovered completely). *Potamogeton* The species of this genus are commonly known as Fishweeds, Pondweeds and River weeds.

Potamogeton Alpinus Balbis: Northern pondweed, Alpine pondweed, Red pondweed Nutrient poor lakes and sloughs in the lowland, steppe and montane zones; frequent throughout BC; circumpolar.

Potamogeton Amplifolius Tucker: Large leaved pondweed, Bass weed, Big pondweed, Broad pondweed, Large leaf pondweed, Muskie weed Lakes and ponds in the lowland and montane zones; frequent in south-western BC, infrequent east of the Coast-Cascade Mountains south of 55 degrees N.

Potamogeton Berchtoldii Fieb. in Bercht: Berchtold's pondweed, Small pondweed Lakes, ponds, sloughs and ditches in the lowland, steppe and montane zones; common throughout BC; circumpolar.

Potamogeton Crispus L: Curled pondweed, Crimped pondweed, Crisp pondweed, Curled pondweed, Curly leaf pondweed, Curly cabbage or muckweed Nutrient rich lakes, ponds and sloughs in the lowland and steppe zones; common in BC south of 50 degrees N; introduced from Eurasia.

Potamogeton Epihydrus Raf: Ribbon leaved pondweed, Leafy pondweed, Nuttall's pondweed Peaty lakes, ditches and ponds from the lowland and steppe to subalpine zones; common in coastal BC, infrequent east of the Coast-Cascade Mountains.

Potamogeton Filiformis Pers: Slender leaved pondweed Calcium rich lakes in the lowland, steppe and montane zones; frequent east of the Coast-Cascade Mountains, rare along the coast; circumpolar.

Potamogeton Foliosus Raf: Closed leaved pondweed, Leafy pondweed, Narrow leaf pondweed Lakes, ponds and ditches in the lowland, steppe and montane zones; frequent in BC south of 55 degrees N, rare in northern BC.

Potamogeton Friesii Rupr: Flat stalked pondweed, Fries' pondweed Lakes, ponds and ditches in the lowland, steppe and montane zones; frequent in BC east of the Coast-Cascade Mountains, infrequent in coastal BC; circumpolar.

Potamogeton Gramineus L: Grass leaved pondweed, Variable leaf pondweed Lakes, lake margins, ponds in peat bogs, ditches and slow flowing streams and rivers in all but the alpine zone; frequent throughout BC; circumpolar. A very plastic and variable species often found growing as a rosette on damp mud when water levels recede.

Potamogeton Illinoensis Morong: Illinois pondweed Lakes and slowly flowing streams in the steppe zone; infrequent in southern BC east of the Coast-Cascade Mountains.

Potamogeton Natans L: Floating leaved pondweed, Bass weed, Broad leaf pondweed, Common floating pondweed, Floating brown leaf, Muskie weed Lakes, streams and ponds from the lowland and steppe to subalpine zones; common throughout BC; cosmopolitan.

Potamogeton Nodosus Poir: Long leaved pondweed, American pondweed, Knotty pondweed, Loddon pondweed, River pondweed, River weed Lakes and sloughs of the lowland zone; infrequent in the lower Fraser Valley, rare elsewhere; cosmopolitan.

Potamogeton Oakesianus Robbins: Oakes' pondweed Lakes in the lowland and steppe zones; rare in southern BC, found in Steelhead and Mara Lakes.

Potamogeton Obtusifolius Mertens and Koch: Blunt leaved pondweed, Grassy pondweed Lakes and sloughs in the lowland, steppe and montane zones; infrequent throughout BC; circumpolar.

Potamogeton Pectinatus L: Sago pondweed, Big pondweed, Large sheath pondweed, Bushy pondweed, Fennel leaved pondweed, Fine leaf pondweed, Giant pondweed, Narrow leaf pondweed, Sheathed pondweed, Slender pondweed, Thread leaf pondweed Lakes, ponds and estuaries in the lowland, steppe and montane zones; frequent in BC east of the Coast-Cascade Mountains, infrequent on the coast; cosmopolitan.

Potamogeton Perfoliatus L: Perfoliate pondweed, Bass weed, Clasping leaf pondweed, Muskie weed, Perfoliate pondweed, Redhead grass Lakes in the montane zone; rare in northern BC, found in Swan Lake in the Cassiar Range; circumpolar.

Potamogeton Praelongus Wulf: Long stalked pondweed, Muskie weed, White stem pond weed Lakes in the lowland, steppe and montane zones; frequent throughout BC; circumpolar.

Potamogeton Pusillus L: Small pondweed, Baby pondweed, Narrow leaf pondweed, Slender leaf pondweed, Small pondweed Lakes, ponds, sloughs and ditches in the lowland, steppe and montane zones; frequent in BC; circumpolar.

Potamogeton Richardsonii (A. Benn.) Rydb: Richardson's pondweed, Clasping leaf pondweed Lakes, ponds and sloughs in the lowland, steppe and montane zones; common throughout BC; amphiberingian.

Potamogeton Robbinsii Oakes: Robbin's pondweed, Fern pondweed, Robbin's pondweed Lakes in the lowland, steppe and montane zones; frequent in coastal BC, infrequent east of the Coast-Cascade Mountains.

Potamogeton Strictifolius Bennett: Stiff leaved pondweed, Narrow leaf pondweed Lakes in the lowland and steppe zones; rare in south-central and south-eastern BC, found in Kawkawa, Mara and Windermere Lakes.

Potamogeton Vaginatus Turcz: Sheathing pondweed Lakes in the montane zone; infrequent east of the Coast-Cascade Mountains; circumpolar.

Potamogeton Zosteriformis Fern: Eel grass pondweed, Flatstem pond weed Lakes in the lowland, steppe and montane zones; frequent in BC east of the Coast-Cascade Mountains, infrequent in coastal BC.

Ranunculus: The species in this genus are commonly known as Buttercups and Crowfoots.

Ranunculus Aquatilis L: White water buttercup, Water crowfoot, Aquatic buttercup Lakes, ponds and slow-moving streams in the lowland, steppe, montane and subalpine zones; frequent throughout BC.

Ranunculus Circinatus Sibth: Stiff leaved water buttercup Ponds and slow-moving streams in the lowland, steppe and montane zones; common in southern BC east of the Coast-Cascade Mountains, less frequent northward.

Ranunculus Cymbalaria Pursh: Shore buttercup, Desert buttercup, Seashore buttercup Moist saline or alkaline shorelines, marshes, wet meadows and shallow marginal water of lakes and ponds in the lowland, steppe and montane zones; common in southern BC, less frequent northward; circumpolar.

Ranunculus Flabellaris Raf: Yellow water buttercup, Yellow crowfoot, Pursh's buttercup Ponds and shallow marginal water of lakes and ponds in the steppe and montane zones; rare and scattered in BC east of the Coast-Cascade Mountains.

Ranunculus Flammula L: Lesser spearwort, Creeping buttercup, Creeping spearwort Wet seepage sites along the shores of lakes, wet meadows and shallow marginal water of lakes and streams in the lowland, steppe and montane zones; common throughout BC; circumpolar.

Ranunculus Gmelinii DC: Small yellow water buttercup Ponds, shallow streams and wet sites in the lowland, steppe and montane zones; common throughout BC except rare along the coast; amphiberingian.

Ranunculus Hyperboreus Rottb: Arctic buttercup, Far northern buttercup, Floating buttercup Ponds and shorelines in the montane zone; frequent in northern BC, rare southward to 53 degrees N; circumpolar.

Ranunculus Lobbii (Hiern) A. Gray: Lobb's water buttercup Vernal pools and wet sites in the lowland zone; rare on south-eastern Vancouver Island, not collected for 50 years.

Ranunculus Sceleratus L: Cursed buttercup, Cursed crowfoot, Celery leaved buttercup, Celery leaved crowfoot Ponds and wet sites in the lowland, steppe and montane zones; common in southern BC east of the Coast-Cascade Mountains, infrequent elsewhere.

Riccia Fluitans L: Crystalwort, Dissected liverwort, Floating liverwort, Floating crystalwort This native aquatic liverwort is rarely collected or reported in aquatic habitats, but is known from a number of scattered sites in BC.

Ricciocarpus Natans (L.) Corda: Purple fringed riccia, Ricciocarpus This native liverwort is rarely collected or reported in aquatic habitats, but is known from a number of scattered sites in BC. (The world distribution is cosmopolitan; the author has seen it growing with *Victoria amazonica* in the Peruvian Amazon basin and noted it in many eutrophic sites throughout BC).

Ruppia Maritima L: Ditch grass, Sea tassel, Widgeon grass Tidal marshes, ponds, lakes and ditches in the lowland, steppe and montane zones; common along the coast and in south-central BC, rare in south-eastern BC; circumpolar. A very variable species or species-complex

Sagittaria: The species in this genus are commonly known as Arrow heads, Duck potatoes, Swamp potatoes and Wapato.

Sagittaria Cuneata Sheldon: Arum leaved arrowhead, Northern arrowhead, Western wapato Wet ditches, ponds and lakes in the lowland, steppe and montane zones; common in BC south of 55 degrees N and east of the Coast-Cascade Mountains, less frequent westward and northward.

Sagittaria Latifolia Willd: Common arrowhead, Broadleaf, Coastal and Slender arrowhead Wet ditches, ponds, lake shores and marshes in the lowland, steppe and montane zones; uncommon in BC south of 56 degrees N, absent from the Queen Charlotte Islands, Northern Vancouver Island and the adjacent coast.

Salvinia: The species in this genus are commonly known as Floating ferns, Floating mosses, Salvinia, Water ferns and Water spangles. Most specific common names are in southeast Asian dialects where the plants are rampant. Introduced tropical free-floating ferns present in the aquarium and water garden trade, occasionally dumped in local lakes but not yet known, or expected, to overwinter in BC. (Although specimens the author has grown in outdoor pools have never overwintered successfully, spores are produced and should be able to survive the winter in suitable habitats.

One species, {*Salvinia natans*}, is known from Europe and is capable of growth in temperate conditions, these plants could become naturalized in BC. {*Salvinia auriculata*} [*Salvinia molesta*] is a tropical species complex, none of which are likely to become established).

Scheuchzeria palustris L: Scheuchzeria Shallow marginal water of lakes and ponds and bogs in the lowland and montane zones; frequent in south-western, central and eastern BC, also along the coast.

Scirpus Lacustris L: American bulrush, Big bulrush, Giant bulrush, Greater bulrush, Porcupine quill bulrush, Slender bulrush, Softstem bulrush, True bulrush, Zebra rush, Zebra bulrush, Tule Shallow water of marshes, lakes and streams in the lowland, steppe and montane zones; common throughout BC. (A common species-complex throughout BC which includes *Scirpus validus* Vahl and *Scirpus acutus* Muhl. in Bigel.).

Scirpus Subterminalis Torr: Water clubrush, Aquatic sedge, Swaying rush Shallow ponds and streams in the lowland zone; infrequent in coastal and south-central BC *Sparganium* The species in this genus are commonly known as Bur reeds.

Sparganium Angustifolium Mich: Narrow leaved bur reed, Floating leaf bur reed, Green fruited bur reed, Slender bur reed, Unbranched bur reed, Western bur reed Ponds, ditches and lake shores in all but the alpine zone; common throughout BC; circumpolar. A highly variable species or species-complex which sometimes includes *Sparganium emersum* as well.

Sparganium Emersum Rehm: Emersed bur reed Streams, ditches, ponds and lake shores in the lowland, steppe and montane zones; common in coastal BC, less frequent elsewhere; circumpolar.

Sparganium Eurycarpum Engelm. in A. Gray: Broad fruited bur reed, Big, Common bur reed, Eastern bur reed, Giant bur reed, Large bur reed, Three square bur reed Wet meadows, shallow ponds and lake shores in the lowland, steppe and montane zones; infrequent in southern and eastern BC.

Sparganium Fluctuans (Morong) B. L. Robins: Water bur reed, Broad ribbon leaf, Floating leaf bur reed, Shining bur reed Ponds, lake shores and slow-moving streams in the lowland and montane zones; rare in south-western and central BC; circumpolar.

Sparganium Glomeratum Laest. ex Beurl: Glomerate bur reed Ponds and lake shores in the lowland and montane zones; rare in west-central BC, known only from Buck Channel, Queen Charlotte Islands and the Smithers area.

Sparganium Hyperboreum Laest. ex Beurl: Northern bur reed Ponds and lake shores in the lowland and montane zones; infrequent throughout BC; circumpolar.

Sparganium Natans L: Small bur reed Ponds, lake shores and slow-moving streams in the lowland and montane zones; common throughout all but north-western BC; circumpolar. [*Sparganium minimum*].

Spirodela Polyrhiza (L.) Schleid: Great duckweed, Big, Greater duckweed, Larger duckweed Ponds, lakes and slow-moving streams in the lowland, steppe and montane zones; common in south-western and south-central BC, infrequent elsewhere, absent on northern Vancouver Island, the Queen Charlotte Islands and north-western BC; circumpolar.

Subularia Aquatica L: Water Awlwort Streams, shorelines, tidally inundated shores, shallow ponds and lakes; infrequent in southern BC.

Tolypella Intricata (Trent.) Leonh: Very little data is available on the abundance and distribution of charophytes in BC.

Trapa Natans L.: Water chestnut, Caltrops, Water nut This species has not yet been found in BC but if introduced it should be able to overwinter and could become a pest.

Utricularia: The species in this genus are commonly known as Bladderworts.

Utricularia Gibba L: Humped bladderwort, Cone spur bladderwort, Eastern bladderwort, Small bladderwort, Yellow bladderwort Lake bottoms and margins, muddy disturbed sites in the lowland zone; rare on southern Vancouver Island and in the Vancouver area.

Utricularia Intermedia Hayne: Flat leaved bladderwort, Mountain bladderwort Oligotrophic and dystrophic lakes and marshes in the lowland, steppe and montane zones; frequent south of 55 degrees N, rare northward; circumpolar.

Utricularia Minor L: Lesser bladderwort, Northern bladderwort, Small bladderwort Oligotrophic and dystrophic lakes and peat bog pools in the lowland and montane zones; common on Vancouver Island and the Queen Charlotte Islands, less frequent inland; circumpolar.

Utricularia Vulgaris L: Greater bladderwort, Common bladderwort Lakes and ponds in the lowland, steppe and montane zones; common south of 55 degrees N in BC, less frequent northwards; circumpolar. *Vallisneria* The species in this genus are commonly known as Eel grasses, Ribbon grasses, Tape grasses and Water celery's.

Vallisneria Americana Michx: American tapegrass Lakes, ponds and streams in the lowland and montane zones; rare in south-western and south-central BC; introduced from eastern North America.

Veronica: The species in this genus are commonly known as Brooklimes and Speedwells.

Veronica Anagallis-aquatica L: Blue water speedwell, Brook pimpernel Wet ditches and, shallow marginal water of streams and lakes in the lowland zone; rare in southern and central BC; introduced from Europe.

Veronica Beccabunga L: American brooklime, American speedwell Wet ditches, stream and lake edges and shallow marginal water in the lowland, steppe and montane zones; common in southern BC, infrequent northward. [*Veronica americana* Schwein.]

Veronica Catenata Pennel: Pink water speedwell, Tufted water speedwell Wet ditches, slow streams, stream and lake edges and shallow marginal water in the lowland and montane zones; rare in southern BC; circumpolar.

Veronica Scutellata L: Marsh speedwell Swamps, stream and lake edges and shallow marginal water in the lowland, steppe and montane zones; frequent throughout BC.

Wolffia: The species in this genus are commonly known as Water meals.

Wolffia Borealis (Engelm. ex Hegelmaier) Landolt and Wildi: Northern water meal, Dotted water meal, Papillary water meal, Spotted water meal, Southern water meal Ponds, lakes and slow-moving streams in the lowland and montane zones; rare in the lower Fraser Valley and south-eastern BC (Creston). This species is probably more common than its collection reports indicate.

Wolffia Columbiana Karsten: Columbian water meal Ponds and lakes in the lowland zone; rare in south-western BC, known only from Beaver, Blenkinsop and Swan Lakes. This species is probably more common than its collection reports indicate.

Wolffiella Floridana (J. D. Smith) Thompson: Florida water meal This species has been tentatively identified by the author from a lake near Smithers but no specimens exist for confirmation.

Zannichellia Palustris L: Horned pondweed, Common poolmat, Grass wrack Shallow, often calcareous or brackish lake shores, slow-moving streams and tidal marshes in the lowland, montane and steppe zones; frequent in mainland BC south of 54 degrees N, less frequent northward and on Vancouver Island; cosmopolitan.

Zostera: The species in this genus are commonly known as Eel grasses.

Zostera Japonica Ascher and Grabn: Japanese eelgrass Tidal mud-flats along the coast in the lowland zone; rare in south-western BC, known only from Boundary Bay and Tsawwassen; introduced from Asia.

Zostera Marina L: Common eelgrass, Wrack Sheltered coastal waters in the lowland zone; common in coastal BC; cosmopolitan.

Appendix: Collecting and Preserving Aquatic Plants

- Introduction
- General Specimen Handling Procedures
- Collecting
- Mounting and Labelling
- Emergent and Wetland Plants
- Small Floating Plants
- Submersed Flaccid and Dissected-leaf Plants
- Plants Requiring Special Handling
- Biomass Studies
- Tissue Analyses.

The collection of aquatic plants is often done as part of ecological or impact studies, both as a record of conditions at a given time for comparison with prior or later conditions, and as a necessity when the collectors are not able to identify the specimens in the field and need to send a specimen to an expert. These specimens are a valuable scientific records and their collection and subsequent handling should be done with care so that the time and expense that has gone into their collection is not wasted.

Herbaria exist for the purpose of long term care and storage of the specimens and uniform standards have been set up for the mounting, care and storage of plant specimens. These standards should be followed so that the plants will become valuable scientific specimens for more than the specific purpose for which they were collected. In addition, their storage in a long term facility like an herbarium will make them accessible to subsequent researchers for other uses. Herbaria provide a permanent record of what was found and if identifications need to be rechecked the specimens are available. Even when the major thrust of the work is for biomass studies or tissue analyses, representative specimens must still be collected and saved as vouchers of what species were analyzed or studied.

There is a good manual published by the BC Ministry of Forests (Aikens, 1996) which is recommended as a reference for collecting, preserving, processing and storing botanical specimens. Most of that information is not repeated here. This appendix is concerned only with the idiosyncrasies of certain aquatic plants which require special handling in order to make good herbarium specimens.

General Specimen Handling Procedures

Plants are mounted on white card stock, about 11 by 16 inches and stored in Herbarium Cabinets made for this purpose. These cabinets have insect-proof tightly-sealed doors, and shelves which accommodate these standard-sized mounting boards. This uniformity facilitates exchange of specimens between institutions and a common design of storage cabinets and facilities. One should assume that a specimen is both permanent and valuable; treat it as such when you collect it, when you are processing and storing it and when you are handling it at a later date. Many specimens in herbaria around the world are hundreds of years old and are of great value in documenting changes in the habitats of areas, and in the study of the ongoing process of plant evolution.

For terrestrial plants, where the plants are rigid and can be mounted dry, most good quality white card stock is acceptable. However most submersed plants do not have structural supporting tissues, they rely on the water for support, and they must be floated onto the card stock. Therefore a paper which remains dimensionally stable after wetting and subsequent drying is required. Most ordinary papers, after being soaked, will wrinkle upon subsequent drying, even if they are kept pressed during the drying process. Ask your paper supplier for a card stock which does not react this way; they are available but their trade names change from time-to-time and supplier-to-supplier.

Collecting

There are a number of aquatic plant collection methods which have been used. The one to use in a given situation depends upon the specific situation and the reason for the collection.

Picking the whole plant by hand is the best method for getting an entire plant in undamaged condition, particularly for preservation as a herbarium specimen. In shallow water this is readily done by wading if the bottom is firm, or by leaning over the side of a canoe where the bottom is too soft for wading. In deeper water snorkeling is the best method. The closer and clearer view of the bottom will often allow one to see small species that were not visible from the surface. In still deeper water SCUBA, or surface-supplied-air diving may be necessary. If there is any wave action, surface ripple or glare, it is difficult to see the bottom from the air; this problem is eliminated when one is underwater.

Rakes or cultivators with long handles can be used to uproot and bring up plants when working from a boat. Cultivators with four long, closely-set prongs are better for uprooting plants but, for small species like Isoetes, it may be difficult to bring the plants to the surface. An Ekman dredge will also bring up small species from deep water if the sediment surface is not too hard. In deep water where visibility is poor one can get a random sample of some of the plants which are present by dragging an anchor from a boat. The plants are not in very good shape and the sample is not necessarily representative of all the species present. Usually all that is collected are fragments of plants without roots or rhizomes and small plants like *Isoetes* will be missed. For sample completeness and specimen quality none of these methods are as good as picking by hand.

Generally the whole plant should be collected. Some groups can not be identified to species without mature fruits or flowers; others need rhizomes, leaf axils or tips. Since submersed aquatic plants do not have to guard against loss of water from their tissues, they do not have waxy or water-repellent cuticles like emergent plants. Do not leave them exposed, even briefly, since they will wither very quickly and become useless as specimens. Keep them in a bag or bucket of water at all times until you are ready to press them. Emergent plants should not be submersed but kept in a bag with a little water in the bottom to maintain a high humidity. It is best to keep each species in its own bag and all the bags from one lake or site together in one large bag.

Some very small plants like the duckweeds do not make very satisfactory pressed and dried specimens, nor is it convenient to collect them into a bag. Small, 20 mL, screw cap vials make good collecting and preservation containers for these plants. Put a little water into the vial to keep the plants moist. Later fill the vial with a solution of 5% formalin, 25% water and 70% ethanol as a permanent preservative. This liquid may need to be replaced after about a month since it will extract chlorophyll and pigments from the plants and become quite dark or opaque. Isopropanol can be used instead of ethanol but ethanol is preferred.

8

Submerged Plants

Submerged plants have stems and leaves that grow entirely underwater, although some may also have floating leaves. Flowers and seeds on short stems that extend above the water may also be present. Submerged plants grow from near shore to the deepest part of the littoral zone and display a wide range of plant shapes. Depending on the species, they may form a low-growing "meadow" near the lake bottom, grow with lots of open space between plant stems, or form dense stands or surface mats.

Broad-leaf Pondweeds *(Potamogeton spp.)*

Common Names: Large-leaf pondweed, claspingleaf (Richardson) pondweed, floating-leaf pondweed, whitestem pondweed, broad-leaf cabbage, musky cabbage, bass weed.

Location: Found in lakes and streams, growing in depths up to 20 feet; often grow near drop-offs.

Description: Leaves 2 to 8 inches long; large-leaf and claspingleaf pondweeds grow below the water surface, except for the flowering stalk; floating-leaf pondweed grows below the water surface, except for their flowering stalk and large, floating leaves; all species have thin and delicate submerged leaves, tough stems that are firmly rooted, and a stiff appearance when out of the water.

Hints to Identify: Leaves alternate along the stem. Often grow in patches or beds; have small seed heads crowded into spikes, often sticking up above water from June through August. Floating-leaf pondweed——Floating leaves are slightly heart-shaped. Large-leaf pondweed——Floating leaves are oval-shaped; submerged leaves are large and wavy; plants are seldom branched. Claspingleaf pondweed—

—Leaves are wide and wavy with a broad base that clasps the stem; plant often branches toward tip; often confused with curlyleaf pondweed, which has small "teeth" along leaf edges, but claspingleaf pondweed has no "teeth" on leaf edges.

Importance of Plants: Broad-leaf pondweeds provide excellent habitat for panfish, largemouth bass, muskellunge, and northern pike; bluegills nest near these plants and eat insects and other small animals found on the leaves; walleyes use these pondweeds for cover.

Management Strategy: These plants are important fish habitat, so it is best to let them be. Removing them may allow less-desirable aquatic plants to move in. These plants are also highly resistant to chemical control. For small problem areas, try raking or cutting the plants.

Bushy Pondweeds and Naiads *(Najas spp.)*

Common Names: Bushy naiad,(pronounced NAY-ads) water naiad, brittle naiad, slender naiad, spiny leaf naiad.

Location: Clear water at depths of up to 20 feet.

Description: Grow entirely below water surface; have long, waving stems in deep water and are dense and bushy in shallow water. These are annual plants which must start from seed each year.

Hints to Identify: Leaves are tapered to a fine point with tiny "spines"; seeds are shiny and smooth. Bushy pondweed is sometimes confused with chara, but chara has a musky odor when crushed and bushy pondweed does not.

Importance of Plants: Entire plants are eaten by waterfowl, especially mallards; provide cover for young largemouth bass, northern pike, small bluegills and perch.

Management Strategy: Bushy pondweed has become more common in recent years; it can grow abundantly in some areas and cause problems. In most areas, it will not be a nuisance and is best left alone—removal may allow less-desirable plants to move in. When control is necessary, aquatic herbicides can be effective.

Canada Waterweed (Elodea canadensis)

Common Names: Elodea (pronounced el-oh-DEE-a), American elodea, common elodea, anacharis, Canada waterweed.

Location: Found in lakes in depths up to 10 feet, often in hard water, and near stream inlets.

Description: Grows entirely underwater, except for a small white flower that blooms during the summer; has branched stems; plants and leaves are usually a dark, grass-green color; leaves are oval-shaped and arranged in clusters of three or four around the stem.

Hints to Identify: Look for leaf clusters compacted near the tip and spaced farther apart down the stem; stems are brittle, branched, and form a large mass near the lake bottom; white flowers have three to four petals and bloom during the summer; flowers are attached to the stem by a threadlike stalk.

Importance of Plant: Provides habitat for many small aquatic animals, which fish and wildlife eat; excellent oxygen producer; attractive leaves make it a popular aquarium plant. However, dense growth of this plant can create a nuisance, and its closed, compact structure is not ideal fish habitat.

Management Strategy: Canada waterweed is often associated with nuisance conditions. If the plant is controlled by cutting or pulling, pick up all fragments—they can grow into new plants. Aquatic herbicides can be an effective control method.

Coontail (Ceratophyllum demersum)

Common Name: Hornwort.

Location: Clear-to-murky water up to 20 feet deep.

Description: Grows underwater with no roots; upper leaves may reach the surface; central hollow stem has stiff, dark-green leaves; plants may be long and sparse, but are often bushy near the tip, giving the plant a "coontail" or "Christmas tree" appearance.

Hints to Identify: Often confused with watermilfoil, but coontail leaves are spiny and forked rather than feather-like.

Importance of Plant: Many waterfowl species eat the shoots; it provides cover for young bluegills, perch, largemouth bass, and northern pike; supports insects that fish and ducklings eat. However, when growing densely, commonly causes nuisance conditions along shorelines.

Management Strategy: Coontail is important to young fish, so remove as little as possible. Use cutting or raking to reduce the amount of plants. Remove all plant fragments from the water because they can regenerate into new plants. Herbicide control can be effective.

Curly-leaf Pondweed (Potamogeton Crispus)

Common Names: Curly cabbage, crisp pondweed.

Location: Grows from the shore to depths of up to 15 feet.

Description: Leaves are somewhat stiff and crinkled, approximately 1/2-inch wide and 2 to 3 inches long; leaves are arranged alternately around the stem, and become more dense toward the end of branches; produces winter buds can be confused with claspingleaf pondweed.

Hints to Identify: Has small "teeth" visible along edge of leaf; begins growing in early spring before most other pondweeds; dies back during midsummer; the flower stalks, when present, stick up above the water surface in June; appears reddish-brown in the water, but is actually green when pulled out of the water and examined closely. Easily confused with claspingleaf pondweed, which has leaves with no "teeth" around their edges.

Importance of Plant: Provides some cover for fish; several waterfowl species feed on the seeds; diving ducks often eat the winter buds.

Management Strategy: Like Eurasian watermilfoil, curlyleaf pondweed is not native to the United States and often causes problems due to excessive growth. When control is necessary, herbicides and harvesting can be effective. Grants are available for control efforts on a lake-wide basis.

Basis for Listing

Potamogeton diversifolius is a small, delicate aquatic plant that occurs in shallow water in small lakes and large ponds.

There are currently only two records of this species from Minnesota. One is a herbarium specimen collected in Ramsey County in 1945 and the other is a herbarium specimen collected in Anoka County in 1992.

Since 1990, over 1,500 lakes in central and northern Minnesota have been surveyed for aquatic plants but this species was not found. Because surveys have been less thorough in the southern part of the state and Minnesota is on the northern limit of this species range, there is still hope it will be found again.

Threats to its aquatic, shallow water habitat continue to grow and include shoreline development, increased recreational activities, and general decline in water quality related to runoff from surrounding uplands. For these reasons, *P. diversifolius* was listed as an endangered species in Minnesota in 1996.

Description

Potamogeton diversifolius closely resembles two other species of pondweeds that also occur in Minnesota, the rare *P. bicupulatus* (snailseed pondweed) and the more common *P. spirillus* (coiled pondweed). It is imperative that good specimens be used for reliable identification and that a specialist verifies the results. In this species, as in *P. bicupulatus*, there are 3 rows of sculpted ridges around the rim of the tiny, disk-shaped seed: 1 ridge on the rim and 2 smaller ridges, 1 on each side, of the rim. *Potamogeton diversifolius* also has extremely fine, narrow leaves but these are slightly wider than *P. bicupulatus*, typically about 0.5 mm (0.02 in.) wide. The measurement must be taken from leaves along typical segments of the underwater stem. Floating leaves, if present, are small and oval, at most about 2 cm (0.8 in) wide and 4 cm (1.6 in.) long, often smaller. Floating leaves are variable in size and are not useful for distinguishing this species (Reznicek and Bobette 1976).

Habitat

There are not enough known locations of *P. diversifolius* to gain a good understanding of its preferred lake habitats. Preliminary indications are that it prefers clear lakes on the Anoka Sand Plain with relatively low levels of dissolved minerals (soft water). These are probably small lakes or large ponds in a gently sloping basin, or perhaps a small, quiet bay in a larger lake.

Conservation / Management

Maintaining high water quality in selected lakes in Minnesota that support this species or other rare aquatic life is a primary challenge. Activities occurring throughout a watershed can affect the water quality of lakes. Runoff can be from adjacent sources, such as shoreline development, or from sources more distant in the watershed, such as industrial manufacturing.

The cumulative impacts on lake water quality and on rare plants such as *P. diversifolius* are not fully known. Certain management activities within a lake itself also pose a possible threat. Aquatic herbicides are widely used, legally or illegally, to kill aquatic vegetation. The target of the herbicides is often invasive, non-native, aquatic plants such as *Myriophyllum spicatum* (Eurasian watermilfoil) or *P. crispus* (curly-leaf pondweed). Herbicides are also used to treat luxuriant growth of native aquatic plant species perceived by lakeshore owners as unsightly or as an obstacle to water access. It is not unusual for these chemicals to be applied with inadequate understanding or consideration of the potential negative impact to rare aquatic species.

Eurasian Watermilfoil (*Myriophyllum Spicatum*)

Eurasian watermilfoil was accidentally introduced to North America from Europe. Spread westward into inland lakes primarily by boats and also by waterbirds, it reached Midwestern states between the 1950s and 1980s.

In nutrient-rich lakes it can form thick underwater stands of tangled stems and vast mats of vegetation at the water's surface. In shallow areas the plant can interfere with water recreation such as boating, fishing, and swimming. The plant's floating canopy can also crowd out important native water plants.

A key factor in the plant's success is its ability to reproduce through stem fragmentation and runners. A single segment of stem and leaves can take root and form a new colony. Fragments clinging to boats and trailers can spread the plant from lake to lake. The mechanical clearing of aquatic plants for beaches, docks, and landings creates thousands of new stem fragments. Removing native vegetation creates perfect habitat for invading Eurasian watermilfoil.

Eurasian watermilfoil has difficulty becoming established in lakes with well established populations of native plants. In some lakes the plant appears to coexist with native flora and has little impact on fish and other aquatic animals.

Likely Means of Spread: Milfoil may become entangled in boat propellers, or may attach to keels and rudders of sailboat. Stems can become lodged among any watercraft apparatus or sports equipment that moves through the water, especially boat trailers.

Narrow-leaf Pondweeds (Potamogeton Zosteriformes, Potamogeton Foliosus, Potamogeton Pectinatus)

Common Names: Flat-stemmed, leafy, and sago pondweeds.

Location: Grow below water surface; are firmly rooted; have branched, slender stems; have grasslike, narrow leaves; may have flowers or seeds extending above water surface; are limp when out of the water.

Hints to Identify: Most narrow-leaf pondweeds have no floating leaves——it is only the seeds of some types that reach or extend above the surface. Sago pondweed has stiff, threadlike leaves that appear bushy and are alternately arranged on one stem; spreading leaves in the water resemble a fan; small, pebble-sized fruits are spaced on the stem and emerge from the water.

Importance of Plants: Provide some cover for bluegills, perch, northern pike, and muskellunge, though these fish prefer broad-leaf pondweeds; good cover for walleye; provide food for waterfowl; support aquatic insects and many other small animals that fish and ducklings eat. Swans, geese, and diving ducks such as canvasbacks favour the tubers and seeds of sago pondweed.

Management Strategy: Removing narrow-leaf pondweeds may allow less-desirable plants such as curlyleaf pondweed or Eurasian watermilfoil to move in. If control is absolutely necessary, try hand-pulling, raking, or underwater cutting. Aquatic herbicides can be an effective control method.

Northern Watermilfoil (Myriophyllum Exalbescens)

Location: Grows entirely underwater in depths from 1 to 20 feet.

Description: Dark-green, feathery leaves are grouped in fours around a hollow stem that is usually buff- or pinkish-colored; leaves are comprised of 5 to 10 pairs of leaflets.

Hints to Identify: Northern watermilfoil is often mistaken for coontail or Eurasian watermilfoil, but it does not branch at the surface as much as Eurasian watermilfoil does; northern typically has half as many leaflet pairs as Eurasian has; northern leaves are rigid when removed from the water, but Eurasian leaves are limp when out of water. The northern species also forms winter buds (groups of small, dark, brittle leaves) in late fall and winter, but the Eurasian variety does not.

Importance of Plant: Provides cover for fish and invertebrates; supports insects and other small animals eaten by fish; waterfowl occasionally eat the fruit and foliage.

Management Strategy: When control is absolutely necessary, hand-pulling, raking, or cutting can be effective in small infestations. Collect and dispose of all fragments——they can regenerate into new plants. Treatment with aquatic herbicides can be effective.

Synonyms

Potamogeton diversifolius var. *trichophyllus*

Basis for Listing

The status of this aquatic plant in Minnesota was unclear for many years because it was known only from a single lake in Anoka

County where it was discovered in 1959. Attempts to locate *Potamogeton bicupulatus* in east-central Minnesota in the late 1980s and early 1990s resulted in discoveries in 4 additional lakes. These same 4 lakes also harbor other rare or uncommon species, possibly indicating that some unique combination of environmental factors is controlling this species' distribution. All of these lakes occur in areas where residential development or other activities are disturbing native vegetation in the surrounding uplands. These findings led to the listing of *P. bicupulatus* as a state endangered species in 1996.

Additional rare plant surveys were conducted in over 1,500 lakes throughout central and north-central Minnesota in the 1990s and early 2000s. *Potamogeton bicupulatus* was found in only 9 more lakes, where it was again associated with other rare or unusual aquatic plants. It tends to occur in lakes that have clear water and relatively low levels of dissolved minerals (soft water lakes). None of the lakes in which it occurs have their surrounding uplands protected, although one site is part of a private conservation area. Residential development of lakeshores continues to be one of the greatest threats to maintaining good water quality over the long-term.

Description

A small, delicate aquatic plant, *P. bicupulatus* closely resembles another rare species of pondweed that also occurs in Minnesota, *P. diversifolius* (diverse-leaved pondweed). All Minnesota sites of *P. bicupulatus* have had fruiting plants present. Fruits, though very small, appear to the naked eye to have a noticeably bumpy surface.

This is due to their 3 rows of sculpted ridges around the rim of a tiny, disk-shaped seed: 1 ridge on the rim and 2 smaller ridges, 1 on each side of the rim. These ridges are also found in *P. diversifolius* , but distinct from most other narrow-leaved pondweeds that have a single ridge or no ridge. *Potamogeton bicupulatus* also has extremely fine, hair-thin leaves, less than 0.3 mm (0.01 in.) wide, along typical segments of the underwater stem. If present, the floating leaves are small and oval, at most about 1 cm (0.4 in.) wide and 2 cm (0.8 in.) long. Floating leaves are variable in size and not definitive in distinguishing this species from other species (Reznicek and Bobette 1976).

Habitat

The characteristics of lake sediments, water quality, and surrounding vegetation in places where *P. bicupulatus* has been found have not been fully analyzed. Preliminary indications are similar to what is observed in the field — that this plant tends to occur in clear, soft water lakes. Typical aquatic plant associates include *Brasenia schreberi* (watershield), *Eriocaulon aquaticum* (pipewort), *Myriophyllum farwellii* (Farwell's watermilfoil), *Myriophyllum tenellum* (slender watermilfoil), *Najas gracillima* (slender naiad), and *Nymphaea odorata* (American white waterlily). Plants have been found rooted in pure sand and in silty sediments along the shores of shallow lakes, and rooted to the bog edge or stranded on low boggy hummocks by receding water.

Biology / Life History

Potamogeton bicupulatus appears to be a perennial herbaceous species that survives winter as a buried rhizome. It apparently does not produce turions, which are vegetative reproductive structures sometimes called winter buds (Haynes and Hellquist 2000).

It is known to produce abundant seeds, which are dispersed on water currents and possibly in the fur or feathers of aquatic animals. Owing to the delicate nature of this plant, it is generally found in calmer portions of larger lakes or in smaller lakes not subject to turbulent wave action. The best time to search for *P. bicupulatus* is when the plant is in fruit during July and August.

Conservation / Management

It is important to maintain high water quality in the select lakes in Minnesota supporting *P. bicupulatus* or assemblages of other rare aquatic plants. Activities occurring throughout a watershed can affect the water quality of lakes. Runoff can be from adjacent sources, such as shoreline development, or from sources more distant in the watershed, such as industrial manufacturing. The cumulative impacts on lake water quality and on rare plants such as *P. bicupulatus* are not fully known. Certain management activities within a lake itself also pose a possible threat. Aquatic herbicides are widely used, legally or illegally, to kill aquatic vegetation. The target of the herbicides is often invasive, non-native, aquatic plants such as *Myriophyllum spicatum* (Eurasian watermilfoil) or *P. crispus* (curly-leaf pondweed).

Herbicides are also used to treat luxuriant growth of native aquatic plant species perceived by lakeshore owners as unsightly or as an obstacle to water access. It is not unusual for these chemicals to be applied with inadequate understanding or consideration of the potential negative impact to rare aquatic species.

Potamogeton bicupulatus has been found consistently in clear-watered lakes with relatively low levels of dissolved minerals (soft water lakes). Therefore activities or actions that alter these characteristics would likely threaten existing plants or prohibit the establishment of new occurrences of the species.

Conservation Efforts in Minnesota

Targeted searches by the Minnesota DNR County Biological Survey for rare aquatic plants in Minnesota are underway. Results of the survey and ongoing water quality monitoring will contribute to our understanding of the relationship between the condition of Minnesota's lakes and rare aquatic vegetation.

Wild Celery *(Vallisneria Americana)*

Common Names: Water celery, eelgrass, tapegrass.

Location: Lakes in depths up to 15 feet and streams; prefers semi-hard bottom such as sand covered with a thin layer of muck.

Description: Leaves are ribbon-like, dark-green, and grow below the water surface; rooted in mud; in late summer, produces a small, whitish-yellow flower, supported by a coiled stalk; often grows in beds amid pondweeds and other submerged plants.

Hints to Identify: Unbranched leaves extending from the lake bottom to the water surface; flowers (and occasionally some leaves) float on the surface; leaves are attached to a horizontal central stem right above lake bottom.

Importance of Plant: Provides shade and shelter for bluegills, young perch, and largemouth bass; choice food of waterfowl, particularly diving ducks; attracts muskrats, marsh birds, and shore birds.

Management Strategy: Because wild celery is an excellent wildlife food, it is usually best left alone. Abundant growth during July and August in shallow water may interfere with recreation. Herbicides don't work well to control this plant. Hand-pulling or raking sometimes works, though floating, uprooted plants often re-establish themselves in shallow water.

Bibliography

Amery, Hussein A.: *Annotated Bibliography of Water in the Middle East*, University of Toronto, Toronto, 1995.

Asaithambi S. : *Economics of Ground Water Management in India*, Abhijeet, Delhi, 2008.

Barman, R P: *Marine and Estuarine Fish Fauna of Orissa*, Zoological Survey of India, Delhi, 2007.

Bullock, John and Darwish, Adel: *Water Wars: Coming Conflicts in the Middle East*, Victor Gollancz, London, 1999.

Chadha K. L. and Pareek O. P.: *Advances in Horticulture: Fruit Crops,* New Delhi, Malhotra Publishing House, 1993.

Cronquist, A.: *The Evolution and Classification of Flowering Plants*, New York Botanical Garden, Bronx, New York., 1988

Diana L.: *People and Pixels: Linking Remote Sensing and Social Science.* National Academies Press, Delhi, 1998.

Doijode S. D.: *Seed Germination in Fruits*, New Delhi, Malhotra Publishers, 1993.

Elachi, Charles : *Introduction to the Physics and Techniques of Remote Sensing*, New York, Wiley, 1987.

Featherly H. I.: *Taxonomic Terminology of the Higher Plants*, USA, Iowa State College Press, 1954.

Ferentinos L.: *Proceeding of the Sustainable Taro Culture for the Pacific Conference*, Honolulu, HITAHR, 1993.

Garrison, T.: *Oceanography, An Invitation to Marine Science*, Wadsworth, Belmont, 1993.

Godfrey,R.K. and Wooten,J.W.: *Aquatic and Wetland Plants of Southeastern United States*, University of Georgia Press, Athens, 1979.

Hammer, D.A. and R.K. Bastian: *Wetlands Ecosystems: Natural Water Purifiers*, Lewis publishers, Chelsea, Michigan, 1989.

Husain, Ahmad : *Environment and Water Resource Management*, Sumit Enterprises, Delhi, 2006.

Johnson, Cait : *Earth, Water, Fire and Air: Essential Ways of Connecting to Spirit*, Woodstock, VT: Skylight Paths, 2003.

Jones, R. M.: *Plant Resources of South-East Asia,* Wageningen, Pudoc Scientific Publishers, 1992.

Kamla Devi and D.V. Rao : *Poisonous and Venomous Fishes of Andaman Islands, Bay of Bengal,* Zoological Survey of India, 2003.

Keddy, P.A.: *Principles and Conservation,* Cambridge University Press, Cambridge, UK, 2010.

Langran, G.: *Time in Geographic Information Systems.* Bristol: Taylor & Francis, 1992.

Lorenzen, S.: *The Phylogenetic Systematics of Freeliving Nematodes*, London, The Ray Society, 1994.

Mason,H.L.: *A flora of the Marshes of California*, University of California Press, Berkeley, 1957.

Matthews, M. C. : *Remote Sensing in Civil Engineering*, Glasgow, Surrey University Press; New York, 1985.

Nobel, P. S.: *Physicochemical and Environmental Plant Physiology*, Academic Press, San Diego, 1999.

Pandey, Prem Chand: *Advances in Marine and Antarctic Science*, APH, Delhi, 2002.

Pemberton, R. W.: *Predictable Risk to Native Plants in Weed Biological Control,* Oecologia, 2000.

Rao K. Nageswara : *Water Resources Management : Realities and Challenges*, New Century Publication, Delhi, 2006.

Rich, T.C.G. and Jermy, A.C.: *Plant Crib 1998*, Botanical Society of the British Isles, London, 1998.

Sahoo, Dinabandhu: *Farming the Ocean : Seaweeds Cultivation and Utilization*, Aravali, Delhi, 2000.

Sculthorpe, C. D.: *The Biology of Aquatic Vascular Plants*, Cambridge University Press, London, 1967.

Todd, David Keith: *Groundwater Hydrology*. John Wiley & Sons, New York, NY. 1959.

Tomlinson, P. B.: *The Botany of Mangroves*, Cambridge University Press, Cambridge, UK, 1986.

Watson, O. Michael : *Symbolic and Expressive use of Space: An Introduction to Proxemic Behavior,* New York, *1972.*

Westlake, D.F., Kvt, J. and Szczepañski, A.: *The Production Ecology of Wetlands*, Cambridge University Press, Cambridge, 1998.

Index

M

N

O

P

S

T

W

❑❑❑